Robotics

Matjaž Mihelj · Tadej Bajd · Aleš Ude
Jadran Lenarčič · Aleš Stanovnik
Marko Munih · Jure Rejc · Sebastjan Šlajpah

Robotics

Second Edition

Springer

Matjaž Mihelj
Faculty of Electrical Engineering
University of Ljubljana
Ljubljana, Slovenia

Tadej Bajd
Faculty of Electrical Engineering
University of Ljubljana
Ljubljana, Slovenia

Aleš Ude
Department of Automatics,
Biocybernetics and Robotics
Jožef Stefan Institute
Ljubljana, Slovenia

Jadran Lenarčič
Jožef Stefan Institute
Ljubljana, Slovenia

Aleš Stanovnik
Faculty of Electrical Engineering
University of Ljubljana
Ljubljana, Slovenia

Marko Munih
Faculty of Electrical Engineering
University of Ljubljana
Ljubljana, Slovenia

Jure Rejc
Faculty of Electrical Engineering
University of Ljubljana
Ljubljana, Slovenia

Sebastjan Šlajpah
Faculty of Electrical Engineering
University of Ljubljana
Ljubljana, Slovenia

ISBN 978-3-030-10285-2 ISBN 978-3-319-72911-4 (eBook)
https://doi.org/10.1007/978-3-319-72911-4

Printed on acid-free paper

This Springer imprint is published by the registered company Springer International Publishing AG part of Springer Nature
The registered company address is: Gewerbestrasse 11, 6330 Cham, Switzerland

Preface

It is perhaps difficult to agree on what a robot is, but most people working in robotics would probably quote the "Father of Robotics", Joseph F. Engelberger (1925–2015), a pioneer in industrial robotics, stating "I can't define a robot, but I know one when I see one".

The word robot does not originate from a scientific or engineering vocabulary, but was first used in the Czech drama "R.U.R." (Rossum's Universal Robots) by Karel Čapek, that was first played in Prague in 1921. The word itself was invented by his brother Josef. In the drama the robot is an artificial human being which is a brilliant worker, deprived of all "unnecessary qualities", such as emotions, creativity, and the capacity for feeling pain. In the prologue of the drama the following definition of robots is given: "Robots are not people (*Roboti nejsou lidé*). They are mechanically more perfect than we are, they have an astounding intellectual capacity, but they have no soul. The creation of an engineer is technically more refined than the product of nature".

The book Robotics evolved through decades of teaching robotics at the Faculty of Electrical Engineering, University of Ljubljana, Slovenia, where the first textbook on industrial robotics was published in 1980 (A. Kralj and T. Bajd, "*Industrijska robotika*"). The way of presenting this rather demanding subject was successfully tested with several generations of undergraduate students.

The second edition of the book continues the legacy of the first edition that won the Outstanding Academic Title distinction from the library magazine CHOICE in 2011. The major feature of the book remains its simplicity. The introductory chapter now comprehensively covers different robot classes with the main focus on industrial robots. The position, orientation, and displacement of an object are described by homogenous transformation matrices. These matrices, which are the basis for any analysis of robot mechanisms, are introduced through simple geometrical reasoning. Geometrical models of the robot mechanism are explained with the help of an original, user-friendly vector description. With the world of the roboticist being six-dimensional, orientation of robot end effectors received more attention in this edition.

Robot kinematics and dynamics are introduced via a mechanism with only two rotational degrees of freedom, which is however an important part of the most popular industrial robot structures. The presentation of robot dynamics is based on only the knowledge of Newton's law and was additionally simplified for easier understanding of this relatively complex matter. The workspace plays an important role in selecting a robot appropriate for the planned task. The kinematics of parallel robots is significantly different from the kinematics of serial manipulators and merits additional attention.

Robot sensors presented in this edition are relevant not only for industrial manipulators, but also for complex systems such as humanoid robots. Robot vision has an increasingly important role in industrial applications and robot trajectory planning is a prerequisite for successful robot control. Basic control schemes, resulting in either the desired end-point trajectory or in the force between the robot and its environment, are explained. Robot environments are illustrated by product assembly processes, where robots are a part of a production line or operate as completely independent units. Robot grippers, tools, and feeding devices are also described.

With the factory floor becoming ever more complex, interaction between humans and robots will be inevitable. Collaborative robots are designed for safe human-robot interaction. Flexibility of production can be further increased with the use of wheeled mobile robots. A glimpse into the future, when humans and robots will be companions, is presented in the chapter on humanoid robotics, the complexity of which requires more advanced knowledge of mathematics. The chapter on standardization and measurement of accuracy and repeatability is of interest for users of industrial robots.

The book requires a minimal advanced knowledge of mathematics and physics. It is therefore appropriate for introductory courses in robotics at engineering faculties (electrical, mechanical, computer, civil). It could also be of interest for engineers who had not studied robotics, but who have encountered robots in the working environment and wish to acquire some basic knowledge in a simple and fast manner.

Ljubljana, Slovenia
April 2018

Matjaž Mihelj
Tadej Bajd

Contents

Chapter 1
Introduction

Today's robotics can be described as a science dealing with intelligent movement of various robot mechanisms which can be classified in the following four groups: robot manipulators, robot vehicles, man-robot systems and biologically inspired robots (Fig. 1.1). The most frequently encountered robot manipulators are serial robot mechanisms. The robot manipulator is represented by a serial chain of rigid bodies, called robot segments, connected by joints. Serial robot manipulators will be described in more details in the next section of this chapter. Parallel robots are of considerable interest both in science and in industry. With these, the robot base and platform are connected to each other with parallel segments, called legs. The segments are equipped with translational actuators, while the joints at the base and platform are passive. Parallel robots are predominantly used for pick-and-place tasks. They are characterized by high accelerations, repeatability, and accuracy. As the robot manipulators replace the human operator at various production jobs, their size is often similar to that of a human arm. Manufacturers can also provide robot manipulators which are up to ten times larger, capable of manipulating complete car bodies. By contrast in the areas of biotechnology and new materials micro- and nanorobots are used. Nanorobots enable pushing, pulling, pick-and-place manipulations, orienting, bending, and grooving on the scale of molecules and particles. The most widespread nanomanipulator is based on the principle of atomic force microscope. The actuator of this nanomanipulator is a piezoelectric crystal, the movement of which is assessed by the use of a laser source and photocell.

Autonomous robot vehicles are found on land, in the water and in the air. The land-based mobile robots are most often applied in man-made environments, such as apartments, hospitals, department stores, or museums, but can increasingly be found on highways and even pathless grounds. Most mobile robots are nevertheless used on flat ground with movement enabled by wheels, with three wheels providing the necessary stability. Often the wheels are specially designed to enable omnidirectional movements. Robot vehicles can be found as vacuum cleaners, autonomous lawn mowers, intelligent guides through department stores or museums, attendants

M. Mihelj et al., *Robotics*, https://doi.org/10.1007/978-3-319-72911-4_1

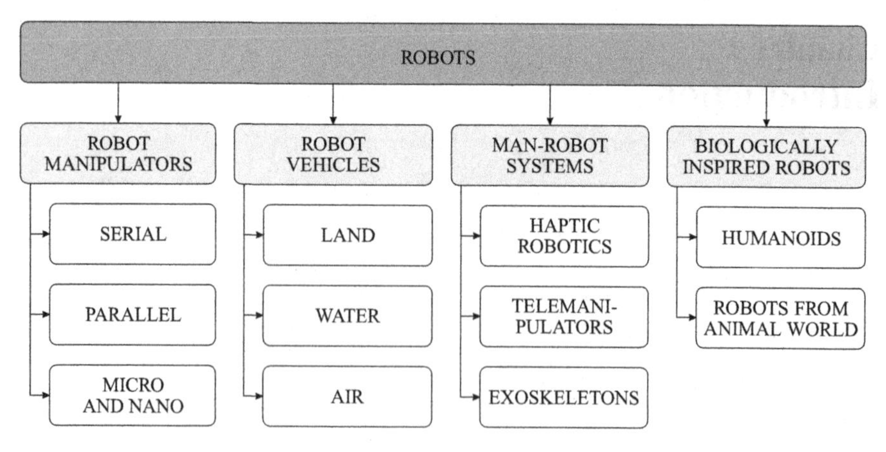

Fig. 1.1 Classification of robots

in clinical centers, space rovers, or autonomous cars. Students can enjoy learning in various competitions, for example football or rescue games, based on the use of small mobile robots. Among the aerial vehicles, the most popular appear to be small quadrocopters. They have a very simple mechanical structure what makes them comparatively inexpensive. Quadrocopters fly using four rotors and are equipped with sensors such as gyroscopes, accelerometers, and cameras, and mostly used for surveillance purposes. Larger autonomous versions are used for military reconnaissance missions. Water-based robots either float on the surface or operate under water. The underwater versions can have the shape of smaller autonomous submarines. They can often be equipped with a robotic arm and used in ocean research, sea floor or ship wrecks observation or as attendants on oil platforms. Autonomous floating robots are used for marine ecological assesments.

New knowledge in the area of robot control is strongly influencing the development of man-robot systems, such as haptic robots, telemanipulators, and exoskeletons. The use of haptic robots is related to virtual environments which are usually displayed on the computer screens. Early virtual environments provided sight and sound to the observer, but no sense of touch. Haptic robots provide the user with the feeling of touch, limited motion, compliance, friction, and texture in virtual environment. Haptic robots play an important role in rehabilitation robotics, where small haptic robots are used for the assessment and evaluation of movements of the upper extremities in paralyzed persons. Stronger haptic systems can hold the wrist of a paralyzed person and guide the arm end-point along the desired path which is shown to the subject in a virtual environment presented on the computer screen. The haptic robot exerts two types of the forces to the subject's wrist. When the patient is unable to perform a movement along the path shown to him in the virtual environment, the robot pushes the wrist along the required trajectory and helps the patient to accomplish the task. The robot is helping only to the extent necessary for the patient to reach the goal point. When the patient's paralyzed extremity travels away from the planned curve, the robot pushes the wrist to the vicinity of the required trajectory. Telemanipulators are robots which are controlled by a human operator when there is a barrier between the telemanipulator and the human operator. The barrier between the operator and working environment is usually either distance (e.g. outer space) or dangerousness (e.g. inside a nuclear plant). Telemanipulators are also entering the medical world, being used in surgery (telemedicine). Exoskeletons are active mechanisms which are attached to human upper or lower extremities. They are mainly used for rehabilitation purposes. Lower limb exoskeletons may increase the strength of healthy persons or enable the retraining of paralyzed persons in walking. In comparison with haptic rehabilitation robots, exoskeletons for upper extremities exert forces to all segments of paralyzed arm.

Biologically inspired robots can be divided into humanoid robots and the robots from the animal world. Examples from the animal world are various types of robotic snakes, fish, quadrupeds, six- or eight leg walking robots. Humanoid robots are by far the most advanced robot systems in the group of the biologically inspired robots. They are designed to live and work in a human environment. The most noticeable property of humanoid robots is their ability of bipedal walking. They walk either with statically stable or dynamically stable gait, they can balance while standing on a single leg, they move in accordance with human co-worker, they can even run. The current problems in humanoid robotics are related to artificial vision, perception and analysis of environment, natural language processing, human interaction, cognitive systems, machine learning and behaviors. Some robots also learn from experience replicating natural processes such as trial-and-error and learning by doing, in the same way a small child learns. In this way the humanoid robot gains a certain degree of autonomy which further means that humanoid robots can behave in some situations in a way that is unpredictable to their human designers. Humanoid robots are coming into our homes and are becoming our partners. They may soon be companions to the elderly and children, assistants to nurses, physicians, firemen, and workers. The need is arising to embody ethics into a robot, which is refered to as robo-ethics. Roboethics is an applied ethics whose objective is to develop scientific/cultural/technical tools that can be shared by different social groups and beliefs. These tools aim to promote and encourage the development of robotics for the advancement of human society and individuals, and to help preventing its misuse against humankind. In 1942 the outstanding novelist Isaac Asimov formulated his famous three laws of robotics. Later, in 1983, he added the fourth law, known as the zeroth law: *No robot may harm humanity or through inaction, allow humanity to come in harm.* The new generation of humanoid robots will be partners that coexist with humans assisting them both physically and psychologically and will contribute to the realization of a safe and peaceful society. They will be potentially more ethical than humans.

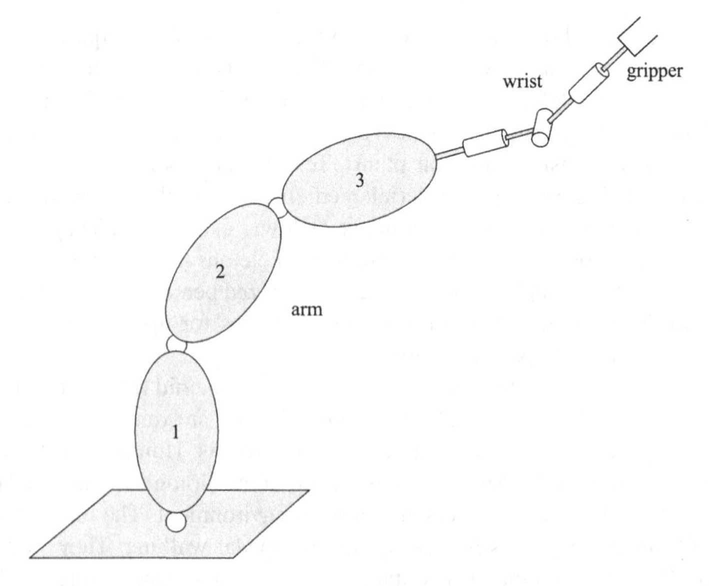

Fig. 1.2 Robot manipulator

1.1 Robot Manipulator

Today the most useful and efficient robotic systems are the industrial robot manipulators which can replace the human workers in difficult or monotonous jobs, or where a human would otherwise be faced with hazardous conditions. The robot manipulator consists of a robot arm, wrist, and gripper (Fig. 1.2). The robot arm is a serial chain of three rigid segments which are relatively long and provide positioning of the gripper in the workspace. Neighboring segments of a robot arm are connected through a robot joint, which is (Fig. 1.3) either translational (prismatic) or rotational (revolute). The rotational joint has the form of a hinge and limits the motion of two neighbor segments to rotation around the joint axis. The relative position is given by the angle of rotation around the joint axis. In robotics the joint angles are denoted by the Greek letter ϑ. In the simplified diagrams the rotational joint is represented by a cylinder. The translational joint restricts the movement of two neighboring segments to translation. The relative position between two segments is measured as a distance. The symbol of the translational joint is a prism, while the distance is denoted by the letter d. Robot joints are powered by either electric or hydraulic motors. The sensors in the joints are measuring the angle or distance, velocity, and torque.

The robot wrist usually consists of three rotational joints. The task of the robot wrist is to enable the required orientation of the object grasped by the robot gripper. The two- or multi-fingered robot gripper is placed at the robot endpoint. Different tools, to enable drilling, spray painting, or welding devices, can be also attached to the endpoint. Industrial robot manipulators usually allow mobility in six degrees of

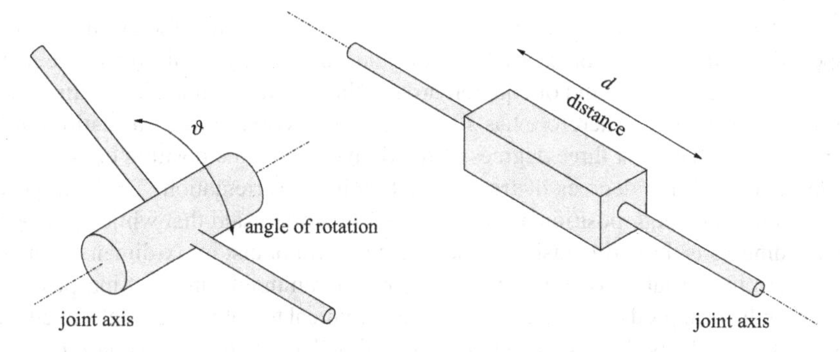

Fig. 1.3 Rotational (left) and translational (right) robot joint

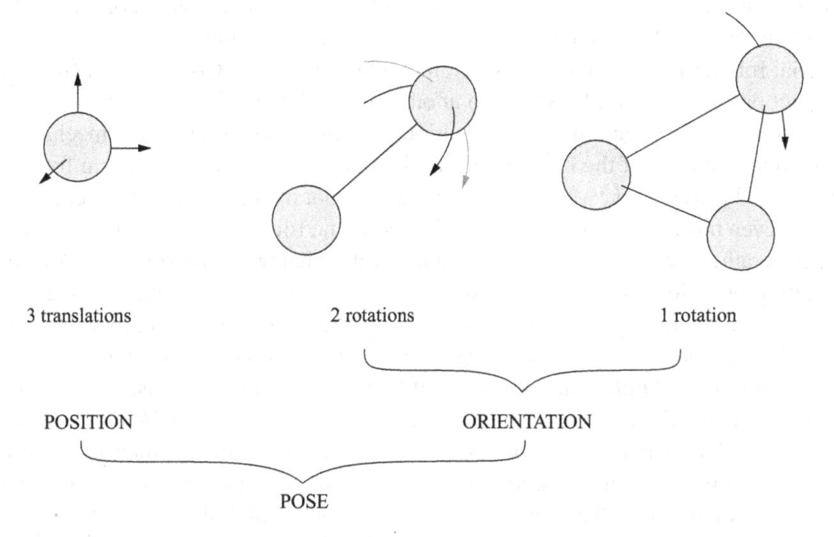

Fig. 1.4 Degrees of freedom of a rigid body

freedom, meaning that the robotic mechanism has six joints and also six actuators. In this way the robot arm can position an object to an arbitrary place in the robot workspace, while the gripper can rotate the object about all three axes of a rectangular coordinate frame.

In order to clarify the term degree of freedom, let us first consider a rigid body which usually represents the object manipulated by the industrial robot. The simplest rigid body consists of three mass particles (Fig. 1.4). A single mass particle has three degrees of freedom, described by three displacements along the axes of a rectangular frame. The displacement along a line is called translation. We add another mass particle to the first one in such a way that there is constant distance between them. The second particle is restricted to move on the surface of a sphere surrounding the first particle. Its position on the sphere can be described by two circles reminding us of meridians and latitudes on a globe. The displacement along a circular line is called rotation. The third mass particle is added in such a way that the distances with respect to the first two particles are kept constant. In this way the third particle may move along the circle, a kind of equator, around the axis determined by the first two particles. A rigid body therefore has six degrees of freedom: three translations and three rotations. The first three degrees of freedom describe the position of the body, while the other three degrees of freedom determine its orientation. The term pose is used to include both position and orientation. It is often said that while the world surrounding us is three-dimensional, the world of a roboticist is six-dimensional.

Modern industrial robot manipulators are reprogrammable and multipurpose. In modern industrial production, it is no longer economical to hold large stocks of either materials or products. This is known as: "Just in time" production. As a consequence, it may happen that different types of a certain product find themselves on the same production line during the same day. This problem, which is most inconvenient for fixed automation devices, can be efficiently resolved by using reprogrammable industrial robotic manipulators. Reprogrammable robots allow us to switch from the production of one type of product to another type by touching a push-button. Furthermore, the robot manipulator is a multipurpose mechanism. The robot mechanism is a crude imitation of the human arm. In the same way as we use our arm for both precise and heavy work, we can apply the same robot manipulator to different tasks. This is even more important in view of the economic life span of an industrial robot, which is rather long (12–16 years). It could therefore happen that a robot manipulator acquired for welding purposes, could be reassigned to a pick and place task. Robot arms have another important property, namely, the axes of two neighboring joints are either parallel or perpendicular. As the robot arm has only three degrees of freedom, there exist a limited number of possible structures of robot arms. Among them the most frequently used are anthropomorphic and the so-called SCARA (Selective Compliant Articulated Robot for Assembly) robot arm. Anthropomorphic type of robot arm (Fig. 1.5), has all three joints of the rotational type, and as such it resembles the human arm to the largest extent. The second joint axis is perpendicular to the first one, while the third joint axis is parallel to the second one. The workspace of the anthropomorphic robot arm, encompassing all the points that can be reached by the robot endpoint, has a spherical shape. The SCARA robot arm appeared relatively late in the development of industrial robotics (Fig. 1.6) and is predominantly used for industrial assembly processes. Two of the joints are rotational and one is translational. The axes of all three joints are parallel. The workspace of SCARA robot arm is of the cylindrical type. In the market we can also find three other commercially available structures of the robot arms: cylindrical, Cartesian, and to a lesser extent spherical.

1.2 Industrial Robotics

Today's industry cannot be imagined any longer without industrial robotic manipulators, which can be divided into three different groups. In the first group we classify the industrial robots which have the role of master in a robot cell. A robot cell usually

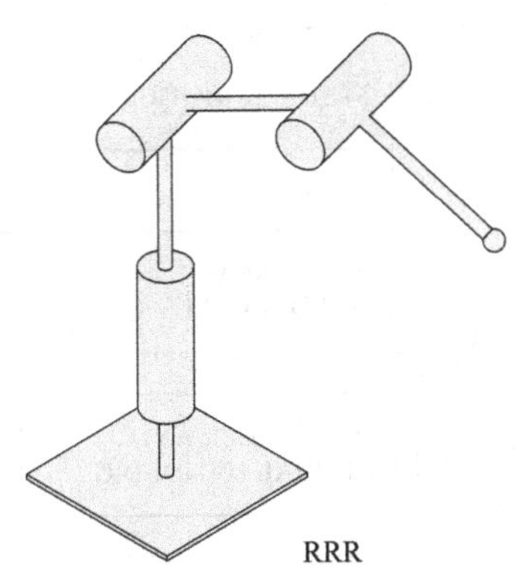

Fig. 1.5 Antropomorphic robot arm

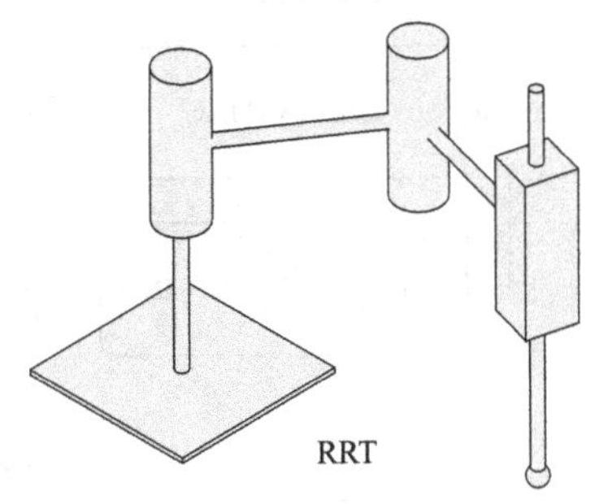

Fig. 1.6 SCARA robot arm

compromises one or more robots, workstations, storage buffers, transport systems and numerically controlled machines. In the second group there are the robots which are slaves within the robot cell. In the third group we include the industrial robots which are used in special applications (Fig. 1.7).

Robot masters in a robot cell, can be found in the following production processes: welding, painting, coating, and sealing, machining, and assembly. Robot welding (spot, arc, laser) represents the most frequent robot applications. It is characterized by speed, precision, and accuracy. Robot welding is specially economic when performed in three shifts. Today we encounter the largest number of welding robots in the car industry. There, the ratio of human workers and robots is 6:1. Industrial robots are often used in aggressive or dangerous environments, such as spray painting. Robotic spray painting represents a saving of material together with a higher quality of painted surfaces. Where toxic environment exist, the social motivation for introduction of robots can outweigh economic factors. In machining applications the robot typically holds either a workpiece or a powered spindle and performs drilling, grinding, deburring or other similar applications. Robot manipulators are increasingly entering the area of industrial assembly, where component parts are assembled into a functional systems. The electronic and electromechanical industries represents

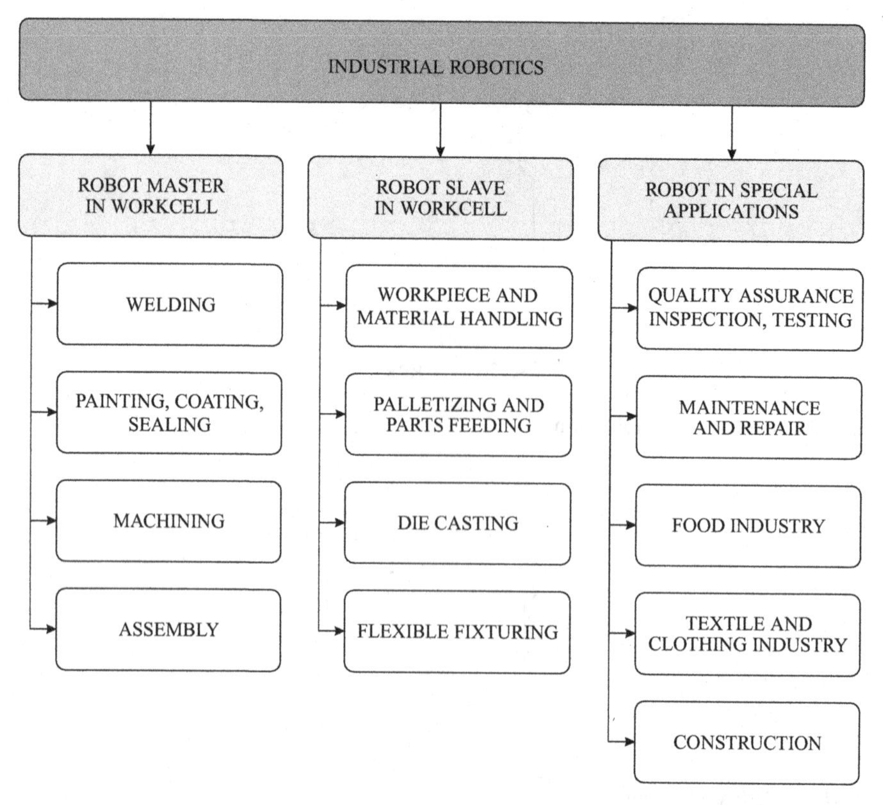

Fig. 1.7 Classification of industrial robots

an important area for the application of assembly robots. There are also attractive assembly operations in the automotive industry, where one robot dispenses adhesive on the windshield glass, while another robot holds the windshield and inserts it into the opening on the vehicle body.

The robot plays the role of a slave in the following industrial applications: workpiece and material handling, palletizing and parts feeding, die casting, and flexible fixturing. In this situation the role of a master can be given to a numerically controlled machine in the robot cell. Pick and place robots represent the most common use of robots in material handling, where tasks are often tedious or repetitive and potentially hazardous (e.g. press loading). Often the industrial robots are used in the tasks when they execute point to point movements. Such examples are encountered in palletizing i.e. arranging of workpieces or products for the purpose of packaging or handing them over to a machine. Robot palletizing is especially appropriate and welcome when heavy objects are considered (e.g. barrels in a brewery). Die casting operations are hot, dirty, and hazardous, providing an unpleasant environment for human workers. With robot handling, the die cast parts are precisely oriented in the die casting machine. The effectiveness of a robot cell can be upgraded by using of flexible fixturing systems. The flexibility of a robot cell is achieved via servodriven programmable positioners, allowing the manufacturing process to be performed faster and more dexterously.

Special applications of the industrial robots are the following: quality assurance, inspection, and testing, maintenance and repair, robots in food, textile and clothing industry, and in construction. Quality assurance, inspection, and testing are often applied in the electronic industry, where electric parameters (e.g. voltage, current, resistance) are tested during assembly of electronic circuits. In this situation the robot performs the necessary measurements on the object (dimensional, electric), while grasping and placing it into a new position. In robot maintenance and repair teleoperated and autonomous robots are used for variety of applications in nuclear industry, highways, railways, power lines maintenance, and aircraft servicing. Robots are also entering the food industry, where in addition to handling and packaging applications in food processing, they are used for the tasks such as food preparation or even decorating chocolates. The textile and clothing industry presents unique problems because of the limp nature of the workpieces, making handling of textiles or similar materials extremely complicated. Many different types of construction robots have been developed all over the world, however very few have been commercialized.

The key challenges of the present-day robotics are human-robot interaction and human-robot collaboration. The development of the so-called soft robotics enables humans and robots to interact and collaborate in industrial environments, in service and everyday settings. When developing collaborative robots, or shortly co-bots, the safety of human-robot interaction must be ensured. Analysis of human injuries caused by blunt or sharp tool impacts was therefore necessary as the first step in collaborative robots research. Based on numerous studies of human-robot collisions, the safe robot velocities were determined for given robot inertial properties. Safe human-robot interaction is further guaranteed by novel control schemes which measure the torque in each robot joint, detecting the slightest contacts between the robot and the human operator and instantly stopping the robot. The prerequisite for the efficient torque control is an extremely detailed model of the dynamics of the robot. To make the robot manipulator compliant, when in contact with human operator, a biologically inspired approach is also used. Storing the energy in the spring elements in the robot arm joints makes the motion control efficient and natural. Complex co-bots, often applied as multi-arm robot systems, cannot be programmed in the same way as ordinary industrial robot manipulators. Cognitive robotics approaches based on artificial intelligence techniques must be introduced, such as imitation learning, learning from demonstrations, reinforcement learning, or learning from rewards. In this way co-bots are able to perform tasks in unknown and unstructured environments. Special attention must also be devoted to robot hands. In collaborating with human operator, the robot hand must be humanoid in order to be able to operate tools and equipment designed for the human hand. Also, the robot hand must measure the forces exerted to provide a gentle grip. The today's industrial robots are for safety reasons still working behind the fences. Fenceless industrial soft robotics has the potential to open novel unforeseen applications, leading to more flexible and cost-effective automation.

Chapter 2
Homogenous Transformation Matrices

2.1 Translational Transformation

As stated previously robots have either translational or rotational joints. To describe the degree of displacement in a joint we need a unified mathematical description of translational and rotational displacements. The translational displacement **d**, given by the vector

$$\mathbf{d} = a\mathbf{i} + b\mathbf{j} + c\mathbf{k}, \tag{2.1}$$

can be described also by the following homogenous transformation matrix **H**

$$\mathbf{H} = Trans(a, b, c) = \begin{bmatrix} 1 & 0 & 0 & a \\ 0 & 1 & 0 & b \\ 0 & 0 & 1 & c \\ 0 & 0 & 0 & 1 \end{bmatrix}. \tag{2.2}$$

When using homogenous transformation matrices an arbitrary vector has the following 4×1 form

$$\mathbf{q} = \begin{bmatrix} x \\ y \\ z \\ 1 \end{bmatrix} = \begin{bmatrix} x \; y \; z \; 1 \end{bmatrix}^T. \tag{2.3}$$

A translational displacement of vector **q** for a distance **d** is obtained by multiplying the vector **q** with the matrix **H**

$$\mathbf{v} = \begin{bmatrix} 1 & 0 & 0 & a \\ 0 & 1 & 0 & b \\ 0 & 0 & 1 & c \\ 0 & 0 & 0 & 1 \end{bmatrix} \begin{bmatrix} x \\ y \\ z \\ 1 \end{bmatrix} = \begin{bmatrix} x + a \\ y + b \\ z + c \\ 1 \end{bmatrix}. \tag{2.4}$$

M. Mihelj et al., *Robotics*, https://doi.org/10.1007/978-3-319-72911-4_2

The translation, which is presented by multiplication with a homogenous matrix, is equivalent to the sum of vectors **q** and **d**

$$\mathbf{v} = \mathbf{q} + \mathbf{d} = (x\mathbf{i} + y\mathbf{j} + z\mathbf{k}) + (a\mathbf{i} + b\mathbf{j} + c\mathbf{k}) = (x + a)\mathbf{i} + (y + b)\mathbf{j} + (z + c)\mathbf{k}. \tag{2.5}$$

In a simple example, the vector $1\mathbf{i} + 2\mathbf{j} + 3\mathbf{k}$ is translationally displaced for the distance $2\mathbf{i} - 5\mathbf{j} + 4\mathbf{k}$

$$\mathbf{v} = \begin{bmatrix} 1 & 0 & 0 & 2 \\ 0 & 1 & 0 & -5 \\ 0 & 0 & 1 & 4 \\ 0 & 0 & 0 & 1 \end{bmatrix} \begin{bmatrix} 1 \\ 2 \\ 3 \\ 1 \end{bmatrix} = \begin{bmatrix} 3 \\ -3 \\ 7 \\ 1 \end{bmatrix}.$$

The same result is obtained by adding the two vectors.

2.2 Rotational Transformation

Rotational displacements will be described in a right-handed rectangular coordinate frame, where the rotations around the three axes, as shown in Fig. 2.1, are considered as positive. Positive rotations around the selected axis are counter-clockwise when looking from the positive end of the axis towards the origin O of the frame x–y–z. The positive rotation can be described also by the so called right hand rule, where the thumb is directed along the axis towards its positive end, while the fingers show the

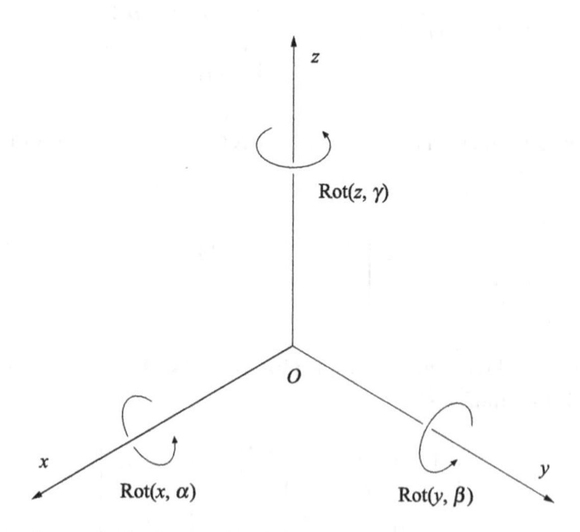

Fig. 2.1 Right-hand rectangular frame with positive rotations

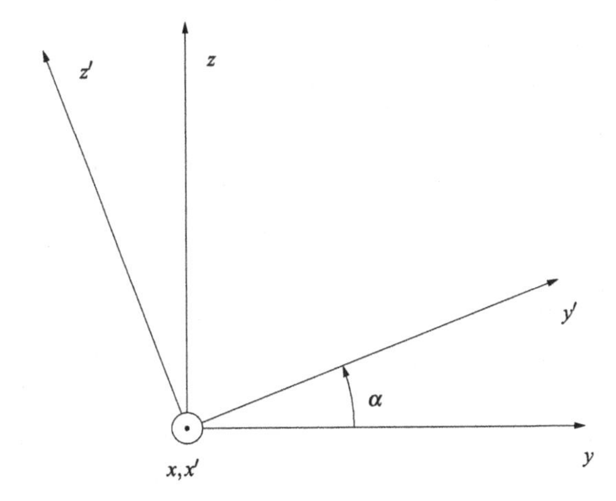

Fig. 2.2 Rotation around x axis

positive direction of the rotational displacement. The direction of running of athletes in a stadium is also an example of a positive rotation.

Let us first take a closer look at the rotation around the x axis. The coordinate frame x'–y'–z' shown in Fig. 2.2 was obtained by rotating the reference frame x–y–z in the positive direction around the x axis for the angle α. The axes x and x' are collinear.

The rotational displacement is also described by a homogenous transformation matrix. The first three rows of the transformation matrix correspond to the x, y, and z axes of the reference frame, while the first three columns refer to the x', y', and z' axes of the rotated frame. The upper left nine elements of the matrix **H** represent the 3×3 rotation matrix. The elements of the rotation matrix are cosines of the angles between the axes given by the corresponding column and row

$$
\begin{aligned}
Rot(x,\alpha) &= \begin{array}{c} \begin{matrix} \quad x' & \quad\quad y' & \quad\quad\quad z' & \end{matrix} \\ \begin{bmatrix} \cos 0^\circ & \cos 90^\circ & \cos 90^\circ & 0 \\ \cos 90^\circ & \cos\alpha & \cos(90^\circ+\alpha) & 0 \\ \cos 90^\circ & \cos(90^\circ-\alpha) & \cos\alpha & 0 \\ 0 & 0 & 0 & 1 \end{bmatrix} \end{array} \begin{matrix} x \\ y \\ z \\ \ \end{matrix} \\
&= \begin{bmatrix} 1 & 0 & 0 & 0 \\ 0 & \cos\alpha & -\sin\alpha & 0 \\ 0 & \sin\alpha & \cos\alpha & 0 \\ 0 & 0 & 0 & 1 \end{bmatrix}.
\end{aligned}
\tag{2.6}
$$

The angle between the x' and the x axes is 0°, therefore we have $\cos 0^\circ$ in the intersection of the x' column and the x row. The angle between the x' and the y axes

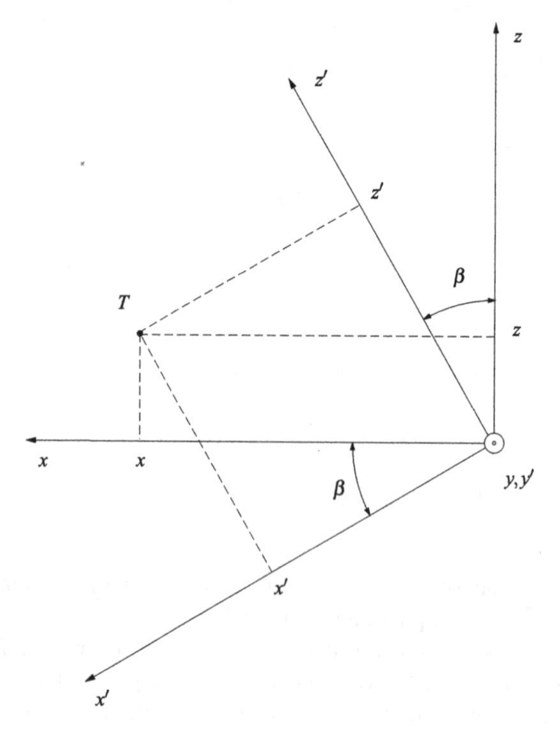

Fig. 2.3 Rotation around y axis

is 90°, we put $\cos 90°$ in the corresponding intersection. The angle between the y' and the y axes is α, the corresponding matrix element is $\cos\alpha$.

To become more familiar with rotation matrices, we shall derive the matrix describing a rotation around the y axis by using Fig. 2.3. The collinear axes are y and y'

$$y = y'. \tag{2.7}$$

By considering the similarity of triangles in Fig. 2.3, it is not difficult to derive the following two equations

$$\begin{aligned} x &= x' \cos\beta + z' \sin\beta \\ z &= -x' \sin\beta + z' \cos\beta. \end{aligned} \tag{2.8}$$

All three Eqs. (2.7) and (2.8) can be rewritten in the matrix form

$$Rot(y,\beta) = \begin{array}{c} \begin{matrix} x' & y' & z' & \end{matrix} \\ \begin{bmatrix} \cos\beta & 0 & \sin\beta & 0 \\ 0 & 1 & 0 & 0 \\ -\sin\beta & 0 & \cos\beta & 0 \\ 0 & 0 & 0 & 1 \end{bmatrix} \end{array} \begin{matrix} x \\ y \\ z \\ \end{matrix}. \tag{2.9}$$

The rotation around the z axis is described by the following homogenous transformation matrix

$$Rot(z,\gamma) = \begin{bmatrix} \cos\gamma & -\sin\gamma & 0 & 0 \\ \sin\gamma & \cos\gamma & 0 & 0 \\ 0 & 0 & 1 & 0 \\ 0 & 0 & 0 & 1 \end{bmatrix}. \tag{2.10}$$

In a simple numerical example we wish to determine the vector $\mathbf{w}$, which is obtained by rotating the vector $\mathbf{u} = 14\mathbf{i} + 6\mathbf{j} + 0\mathbf{k}$ for 90° in the counter clockwise (i.e., positive) direction around the z axis. As $\cos 90° = 0$ and $\sin 90° = 1$, it is not difficult to determine the matrix describing $Rot(z, 90°)$ and multiplying it by the vector $\mathbf{u}$

$$\mathbf{w} = \begin{bmatrix} 0 & -1 & 0 & 0 \\ 1 & 0 & 0 & 0 \\ 0 & 0 & 1 & 0 \\ 0 & 0 & 0 & 1 \end{bmatrix} \begin{bmatrix} 14 \\ 6 \\ 0 \\ 1 \end{bmatrix} = \begin{bmatrix} -6 \\ 14 \\ 0 \\ 1 \end{bmatrix}.$$

The graphical presentation of rotating the vector $\mathbf{u}$ around the z axis is shown in Fig. 2.4.

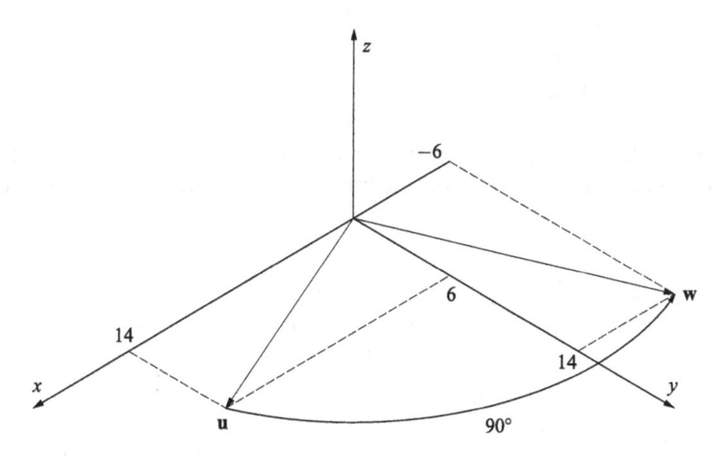

Fig. 2.4 Example of rotational transformation

2.3 Pose and Displacement

In the previous section we have learned how a point is translated or rotated around the axes of the cartesian frame. In continuation we shall be interested in displacements of objects. We can always attach a coordinate frame to a rigid object under consideration. In this section we shall deal with the pose and the displacement of rectangular frames. Here we see that a homogenous transformation matrix describes either the pose of a frame with respect to a reference frame, or it represents the displacement of a frame into a new pose. In the first case the upper left 3×3 matrix represents the orientation of the object, while the right-hand 3×1 column describes its position (e.g., the position of its center of mass). The last row of the homogenous transformation matrix will be always represented by [0 0 0 1]. In the case of object displacement, the upper left matrix corresponds to rotation and the right-hand column corresponds to translation of the object. We shall examine both cases through simple examples. Let us first clear up the meaning of the homogenous transformation matrix describing the pose of an arbitrary frame with respect to the reference frame. Let us consider the following product of homogenous matrices which gives a new homogenous transformation matrix **H**

$$\begin{aligned}\mathbf{H} &= Trans(8,-6,14)Rot(y,90^\circ)Rot(z,90^\circ)\\ &= \begin{bmatrix}1&0&0&8\\0&1&0&-6\\0&0&1&14\\0&0&0&1\end{bmatrix}\begin{bmatrix}0&0&1&0\\0&1&0&0\\-1&0&0&0\\0&0&0&1\end{bmatrix}\begin{bmatrix}0&-1&0&0\\1&0&0&0\\0&0&1&0\\0&0&0&1\end{bmatrix}\\ &= \begin{bmatrix}0&0&1&8\\1&0&0&-6\\0&1&0&14\\0&0&0&1\end{bmatrix}.\end{aligned} \tag{2.11}$$

When defining the homogenous matrix representing rotation, we learned that the first three columns describe the rotation of the frame x'–y'–z' with respect to the reference frame x–y–z

$$\begin{array}{c} \quad x' \quad y' \quad z' \\ \begin{bmatrix}\begin{bmatrix}0\\1\\0\end{bmatrix}&\begin{bmatrix}0\\0\\1\end{bmatrix}&\begin{bmatrix}1\\0\\0\end{bmatrix}&\begin{matrix}8\\-6\\14\end{matrix}\\0&0&0&1\end{bmatrix}\begin{matrix}x\\y\\z\\ \end{matrix}\end{array}. \tag{2.12}$$

The fourth column represents the position of the origin of the frame x'–y'–z' with respect to the reference frame x–y–z. With this knowledge we can represent graphically the frame x'–y'–z' described by the homogenous transformation matrix (2.11), relative to the reference frame x–y–z (Fig. 2.5). The x' axis points in the

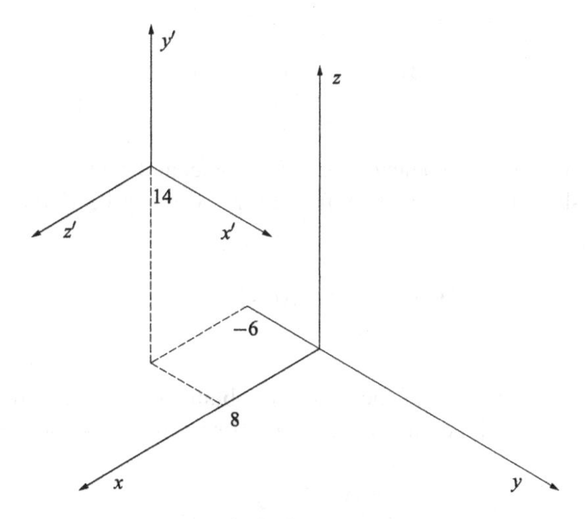

Fig. 2.5 The pose of an arbitrary frame x'–y'–z' with respect to the reference frame x–y–z

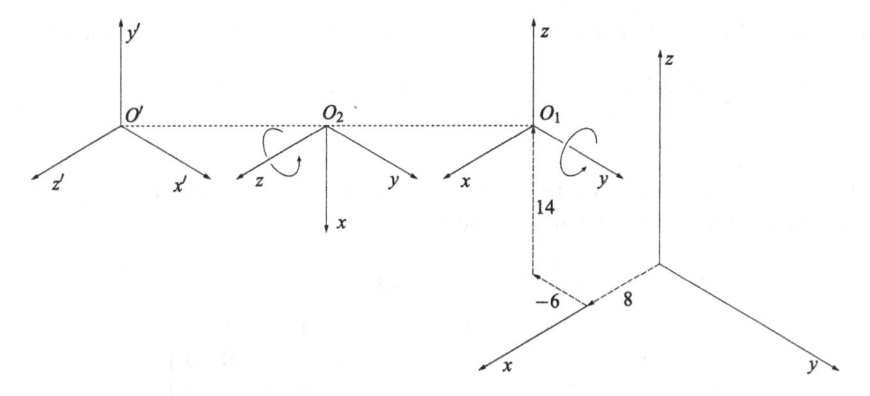

Fig. 2.6 Displacement of the reference frame into a new pose (from right to left). The origins O_1, O_2 and O' are in the same point

direction of y axis of the reference frame, the y' axis is in the direction of the z axis, and the z' axis is in the x direction.

To convince ourselves of the correctness of the frame drawn in Fig. 2.6, we shall check the displacements included in Eq. (2.11). The reference frame is first translated into the point (8, −6, 14), afterwards it is rotated for 90° around the new y axis and finally it is rotated for 90° around the newest z axis (Fig. 2.6). The three displacements of the reference frame result in the same final pose as shown in Fig. 2.5.

In continuation of this chapter we wish to elucidate the second meaning of the homogenous transformation matrix, i.e., a displacement of an object or coordinate frame into a new pose (Fig. 2.7). First, we wish to rotate the coordinate frame x–y–z for 90° in the counter-clockwise direction around the z axis. This can be achieved by the following post-multiplication of the matrix **H** describing the initial pose of the coordinate frame x–y–z

$$\mathbf{H}_1 = \mathbf{H} \cdot Rot(z, 90^\circ). \tag{2.13}$$

The displacement resulted in a new pose of the object and new frame x'–y'–z' shown in Fig. 2.7. We shall displace this new frame for -1 along the x' axis, 3 units along y' axis and -3 along z' axis

$$\mathbf{H}_2 = \mathbf{H}_1 \cdot Trans(-1, 3, -3). \tag{2.14}$$

After translation a new pose of the object is obtained together with a new frame x''–y''–z''. This frame will be finally rotated for 90° around the y'' axis in the positive direction

$$\mathbf{H}_3 = \mathbf{H}_2 \cdot Rot(y'', 90^\circ). \tag{2.15}$$

The Eqs. (2.13), (2.14), and (2.15) can be successively inserted one into another

$$\mathbf{H}_3 = \mathbf{H} \cdot Rot(z, 90^\circ) \cdot Trans(-1, 3, -3) \cdot Rot(y'', 90^\circ) = \mathbf{H} \cdot \mathbf{D}. \tag{2.16}$$

In Eq. (2.16), the matrix $\mathbf{H}$ represents the initial pose of the frame, $\mathbf{H}_3$ is the final pose, while $\mathbf{D}$ represents the displacement

$$\begin{aligned} \mathbf{D} &= Rot(z, 90^\circ) \cdot Trans(-1, 3, -3) \cdot Rot(y'', 90^\circ) \\ &= \begin{bmatrix} 0 & -1 & 0 & 0 \\ 1 & 0 & 0 & 0 \\ 0 & 0 & 1 & 0 \\ 0 & 0 & 0 & 1 \end{bmatrix} \begin{bmatrix} 1 & 0 & 0 & -1 \\ 0 & 1 & 0 & 3 \\ 0 & 0 & 1 & -3 \\ 0 & 0 & 0 & 1 \end{bmatrix} \begin{bmatrix} 0 & 0 & 1 & 0 \\ 0 & 1 & 0 & 0 \\ -1 & 0 & 0 & 0 \\ 0 & 0 & 0 & 1 \end{bmatrix} \\ &= \begin{bmatrix} 0 & -1 & 0 & -3 \\ 0 & 0 & 1 & -1 \\ -1 & 0 & 0 & -3 \\ 0 & 0 & 0 & 1 \end{bmatrix}. \end{aligned} \tag{2.17}$$

Finally, we shall perform the post-multiplication describing the new relative pose of the object

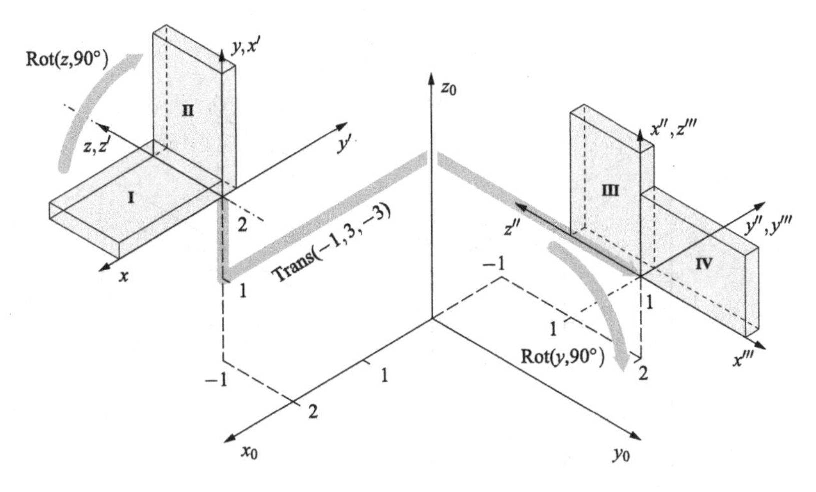

Fig. 2.7 Displacement of the object into a new pose

$$
\mathbf{H}_3 = \mathbf{H} \cdot \mathbf{D} = \begin{bmatrix} 1 & 0 & 0 & 2 \\ 0 & 0 & -1 & -1 \\ 0 & 1 & 0 & 2 \\ 0 & 0 & 0 & 1 \end{bmatrix} \begin{bmatrix} 0 & -1 & 0 & -3 \\ 0 & 0 & 1 & -1 \\ -1 & 0 & 0 & -3 \\ 0 & 0 & 0 & 1 \end{bmatrix}
$$
$$
\begin{matrix} \quad x''' & y''' & z''' & \end{matrix}
$$
$$
= \begin{bmatrix} 0 & -1 & 0 & -1 \\ 1 & 0 & 0 & 2 \\ 0 & 0 & 1 & 1 \\ 0 & 0 & 0 & 1 \end{bmatrix} \begin{matrix} x_0 \\ y_0 \\ z_0 \\ \end{matrix} . \tag{2.18}
$$

As in the previous example we shall graphically verify the correctness of the matrix (2.18). The three displacements of the frame x–y–z: rotation for 90° in counterclockwise direction around the z axis, translation for -1 along the x' axis, 3 units along y' axis and -3 along z' axis, and rotation for 90° around y'' axis in the positive direction are shown in Fig. 2.7. The result is the final pose of the object x''', y''', z'''. The x''' axis points in the positive direction of the y_0 axis, y''' points in the negative direction of x_0 axis and z''' points in the positive direction of z_0 axis of the reference frame. The directions of the axes of the final frame correspond to the first three columns of the matrix $\mathbf{H}_3$. There is also agreement between the position of the origin of the final frame in Fig. 2.7 and the fourth column of the matrix $\mathbf{H}_3$.

2.4 Geometrical Robot Model

Our final goal is the geometrical model of a robot manipulator. A geometrical robot model is given by the description of the pose of the last segment of the robot (end-effector) expressed in the reference (base) frame. The knowledge how to describe the

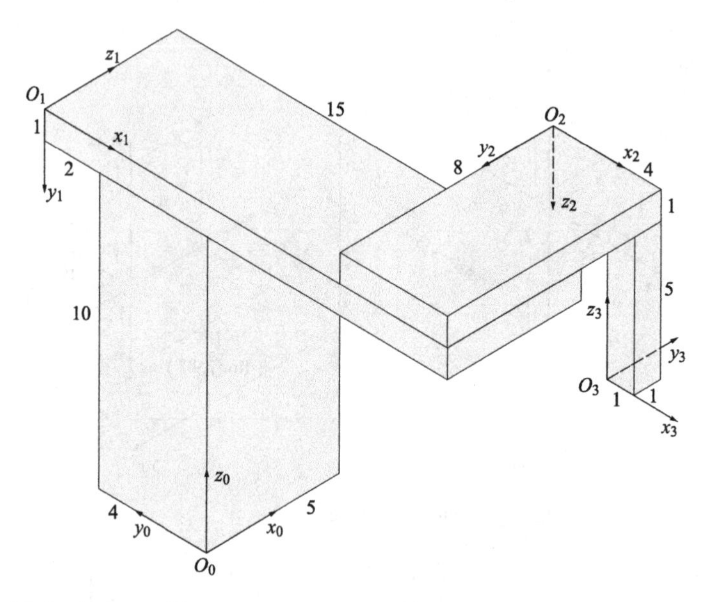

Fig. 2.8 Mechanical assembly

pose of an object using homogenous transformation matrices will be first applied to the process of assembly. For this purpose, a mechanical assembly consisting of four blocks, such as presented in Fig. 2.8, will be considered. A plate with dimensions ($5 \times 15 \times 1$) is placed over a block ($5 \times 4 \times 10$). Another plate ($8 \times 4 \times 1$) is positioned perpendicularly to the first one, holding another small block ($1 \times 1 \times 5$).

A frame is attached to each of the four blocks as shown in Fig. 2.8. Our task will be to calculate the pose of the frame x_3–y_3–z_3 with respect to the reference frame x_0–y_0–z_0. In the last chapter we learned that the pose of a displaced frame can be expressed with respect to the reference frame using the homogenous transformation matrix $\mathbf{H}$. The pose of the frame x_1–y_1–z_1 with respect to the frame x_0–y_0–z_0 will be denoted by ${}^0\mathbf{H}_1$. In the same way ${}^1\mathbf{H}_2$ represents the pose of frame x_2–y_2–z_2 with respect to x_1–y_1–z_1 and ${}^2\mathbf{H}_3$ the pose of x_3–y_3–z_3 with regard to frame x_2–y_2–z_2. We learned also that the successive displacements are expressed by post-multiplications (successive multiplications from left to right) of homogenous transformation matrices. The assembly process can be described by post-multiplication of the corresponding matrices. The pose of the fourth block can be written with respect to the first one by the following matrix

$$
{}^0\mathbf{H}_3 = {}^0\mathbf{H}_1{}^1\mathbf{H}_2{}^2\mathbf{H}_3. \tag{2.19}
$$

The blocks were positioned perpendicularly one to another. In this way it is not necessary to calculate the sines and cosines of the rotation angles. The matrices can be determined directly from Fig. 2.8. The x axis of frame x_1–y_1–z_1 points in negative direction of the y axis in the frame x_0–y_0–z_0. The y axis of frame x_1–y_1–z_1 points in negative direction of the z axis in the frame x_0–y_0–z_0. The z axis of the frame x_1–y_1–z_1 has the same direction as x axis of the frame x_0–y_0–z_0. The described geometrical properties of the assembly structure are written into the first three columns of the homogenous matrix. The position of the origin of the frame x_1–y_1–z_1 with respect to the frame x_0–y_0–z_0 is written into the fourth column

$$
{}^0\mathbf{H}_1 = \overset{\overbrace{x \quad y \quad z}^{O_1}}{\begin{bmatrix} 0 & 0 & 1 & 0 \\ -1 & 0 & 0 & 6 \\ 0 & -1 & 0 & 11 \\ 0 & 0 & 0 & 1 \end{bmatrix}} \left.\begin{matrix} x \\ y \\ z \end{matrix}\right\} O_0 . \tag{2.20}
$$

In the same way the other two matrices are determined

$$
{}^1\mathbf{H}_2 = \begin{bmatrix} 1 & 0 & 0 & 11 \\ 0 & 0 & 1 & -1 \\ 0 & -1 & 0 & 8 \\ 0 & 0 & 0 & 1 \end{bmatrix} \tag{2.21}
$$

$$
{}^2\mathbf{H}_3 = \begin{bmatrix} 1 & 0 & 0 & 3 \\ 0 & -1 & 0 & 1 \\ 0 & 0 & -1 & 6 \\ 0 & 0 & 0 & 1 \end{bmatrix} . \tag{2.22}
$$

The position and orientation of the fourth block with respect to the first one is given by the ${}^0\mathbf{H}_3$ matrix which is obtained by successive multiplication of the matrices (2.20), (2.21) and (2.22)

$$
{}^0\mathbf{H}_3 = \begin{bmatrix} 0 & 1 & 0 & 7 \\ -1 & 0 & 0 & -8 \\ 0 & 0 & 1 & 6 \\ 0 & 0 & 0 & 1 \end{bmatrix} . \tag{2.23}
$$

The fourth column of the matrix ${}^0\mathbf{H}_3$ $[7, -8, 6, 1]^T$ represents the position of the origin of the frame x_3–y_3–z_3 with respect to the reference frame x_0–y_0–z_0. The accuracy of the fourth column can be checked from Fig. 2.8. The rotational part of the matrix ${}^0\mathbf{H}_3$ represents the orientation of the frame x_3–y_3–z_3 with respect to the reference frame x_0–y_0–z_0.

Now let us imagine that the first horizontal plate rotates with respect to the first vertical block around axis 1 for angle ϑ_1. The second plate also rotates around the vertical axis 2 for angle ϑ_2. The last block is elongated for distance d_3 along the third

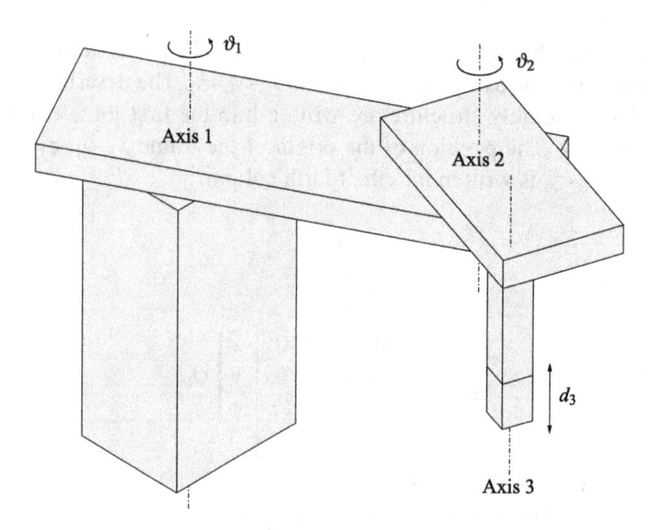

Fig. 2.9 Displacements of the mechanical assembly

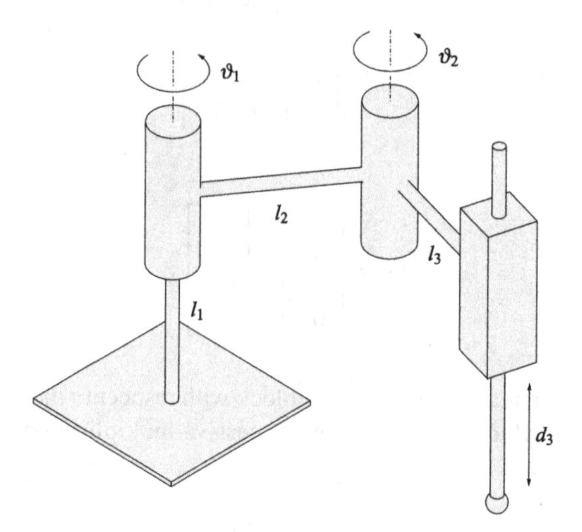

Fig. 2.10 SCARA robot manipulator in an arbitrary pose

axis. In this way we obtained a robot manipulator, of the SCARA type as mentioned in the introductory chapter.

Our goal is to develop a geometrical model of the SCARA robot. Blocks and plates from Fig. 2.9 will be replaced by symbols for rotational and translational joints that we know from the introduction (Fig. 2.10).

The first vertical segment with the length l_1 starts from the base (where the robot is attached to the ground) and is terminated by the first rotational joint. The second segment with length l_2 is horizontal and rotates around the first segment. The rotation in the first joint is denoted by the angle ϑ_1. The third segment with the length l_3 is also

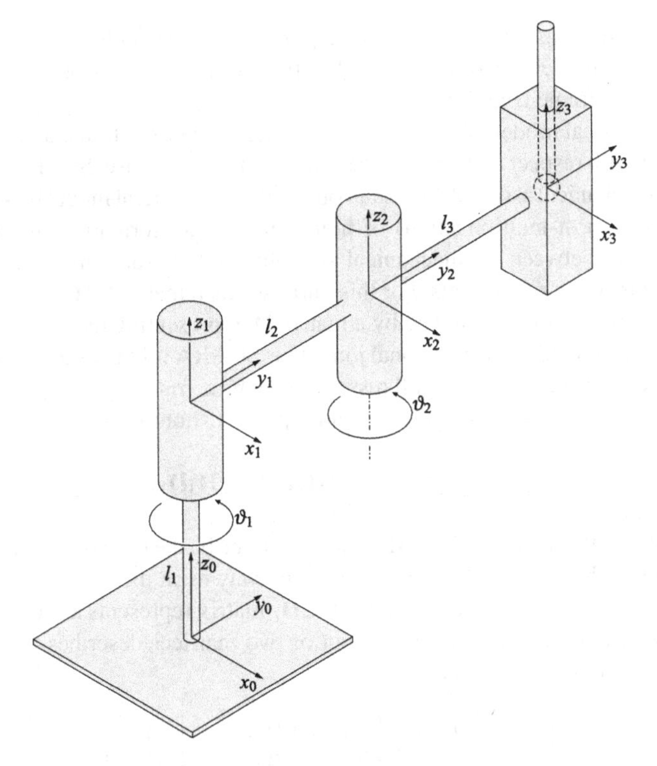

Fig. 2.11 The SCARA robot manipulator in the initial pose

horizontal and rotates around the vertical axis at the end of the second segment. The angle is denoted as ϑ_2. There is a translational joint at the end of the third segment. It enables the robot end-effector to approach the working plane where the robot task takes place. The translational joint is displaced from zero initial length to the length described by the variable d_3.

The robot mechanism is first brought to the initial pose which is also called "home position". In the initial pose two neighboring segments must be either parallel or perpendicular. The translational joints are in their initial position $d_i = 0$. The initial pose of the SCARA manipulator is shown in Fig. 2.11.

First, the coordinate frames must be drawn into the SCARA robot presented in Fig. 2.11. The first (reference) coordinate frame x_0–y_0–z_0 is placed onto the base of the robot. In the last chapter we shall learn that robot standards require the z_0 axis to point perpendicularly out from the base. In this case it is aligned with the first segment. The other two axes are selected in such a way that robot segments are parallel to one of the axes of the reference coordinate frame, when the robot is in its initial home position. In this case we align the y_0 axis with the segments l_2 and l_3. The coordinate frame must be right handed. The rest of the frames are placed into the robot joints. The origins of the frames are drawn in the center of each joint. One of the frame axes must be aligned with the joint axis. The simplest way to calculate the geometrical model of a robot is to make all the frames in the robot joints parallel to the reference frame (Fig. 2.11).

The geometrical model of a robot describes the pose of the frame attached to the end-effector with respect to the reference frame on the robot base. Similarly, as in the case of the mechanical assembly, we shall obtain the geometrical model by successive multiplication (post-multiplication) of homogenous transformation matrices. The main difference between the mechanical assembly and the robot manipulator is the displacements of the robot joints. For this purpose, each matrix ${}^{i-1}\mathbf{H}_i$ describing the pose of a segment will be followed by a matrix $\mathbf{D}_i$ representing the displacement of either the translational or the rotational joint. Our SCARA robot has three joints. The pose of the end frame x_3–y_3–z_3 with respect to the base frame x_0–y_0–z_0 is expressed by the following postmultiplication of three pairs of homogenous transformation matrices

$$ {}^{0}\mathbf{H}_3 = ({}^{0}\mathbf{H}_1\mathbf{D}_1)\cdot({}^{1}\mathbf{H}_2\mathbf{D}_2)\cdot({}^{2}\mathbf{H}_3\mathbf{D}_3). \tag{2.24} $$

In Eq. (2.24), the matrices ${}^{0}\mathbf{H}_1$, ${}^{1}\mathbf{H}_2$, and ${}^{2}\mathbf{H}_3$ describe the pose of each joint frame with respect to the preceding frame in the same way as in the case of assembly of the blocs. From Fig. 2.11 it is evident that the $\mathbf{D}_1$ matrix represents a rotation around the positive z_1 axis. The following product of two matrices describes the pose and the displacement in the first joint

$$ {}^{0}\mathbf{H}_1\mathbf{D}_1 = \begin{bmatrix} 1 & 0 & 0 & 0 \\ 0 & 1 & 0 & 0 \\ 0 & 0 & 1 & l_1 \\ 0 & 0 & 0 & 1 \end{bmatrix} \begin{bmatrix} c1 & -s1 & 0 & 0 \\ s1 & c1 & 0 & 0 \\ 0 & 0 & 1 & 0 \\ 0 & 0 & 0 & 1 \end{bmatrix} = \begin{bmatrix} c1 & -s1 & 0 & 0 \\ s1 & c1 & 0 & 0 \\ 0 & 0 & 1 & l_1 \\ 0 & 0 & 0 & 1 \end{bmatrix}. $$

In the above matrices the following shorter notation was used $\sin\vartheta_1 = s1$ and $\cos\vartheta_1 = c1$.

In the second joint there is a rotation around the z_2 axis

$$ {}^{1}\mathbf{H}_2\mathbf{D}_2 = \begin{bmatrix} 1 & 0 & 0 & 0 \\ 0 & 1 & 0 & l_2 \\ 0 & 0 & 1 & 0 \\ 0 & 0 & 0 & 1 \end{bmatrix} \begin{bmatrix} c2 & -s2 & 0 & 0 \\ s2 & c2 & 0 & 0 \\ 0 & 0 & 1 & 0 \\ 0 & 0 & 0 & 1 \end{bmatrix} = \begin{bmatrix} c2 & -s2 & 0 & 0 \\ s2 & c2 & 0 & l_2 \\ 0 & 0 & 1 & 0 \\ 0 & 0 & 0 & 1 \end{bmatrix}. $$

In the last joint there is translation along the z_3 axis

$$ {}^{2}\mathbf{H}_3\mathbf{D}_3 = \begin{bmatrix} 1 & 0 & 0 & 0 \\ 0 & 1 & 0 & l_3 \\ 0 & 0 & 1 & 0 \\ 0 & 0 & 0 & 1 \end{bmatrix} \begin{bmatrix} 1 & 0 & 0 & 0 \\ 0 & 1 & 0 & 0 \\ 0 & 0 & 1 & -d_3 \\ 0 & 0 & 0 & 1 \end{bmatrix} = \begin{bmatrix} 1 & 0 & 0 & 0 \\ 0 & 1 & 0 & l_3 \\ 0 & 0 & 1 & -d_3 \\ 0 & 0 & 0 & 1 \end{bmatrix}. $$

The geometrical model of the SCARA robot manipulator is obtained by post-multiplication of the three matrices derived above

$$
{}^{0}\mathbf{H}_3 = \begin{bmatrix} c12 & -s12 & 0 & -l_3 s12 - l_2 s1 \\ s12 & c12 & 0 & l_3 c12 + l_2 c1 \\ 0 & 0 & 1 & l_1 - d_3 \\ 0 & 0 & 0 & 1 \end{bmatrix}.
$$

When multiplying the three matrices the following abbreviation was introduced $c12 = \cos(\vartheta_1 + \vartheta_2) = c1c2 - s1s2$ and $s12 = \sin(\vartheta_1 + \vartheta_2) = s1c2 + c1s2$.

Chapter 3
Geometric Description of the Robot Mechanism

The geometric description of the robot mechanism is based on the usage of translational and rotational homogenous transformation matrices. A coordinate frame is attached to the robot base and to each segment of the mechanism, as shown in Fig. 3.1. Then, the corresponding transformation matrices between the consecutive frames are determined. A vector expressed in one of the frames can be transformed into another frame by successive multiplication of intermediate transformation matrices.

Vector **a** in Fig. 3.1 is expressed relative to the coordinate frame x_3–y_3–z_3, while vector **b** is given in the frame x_0–y_0–z_0 belonging to the robot base. A mathematical relationship between the two vectors is obtained by the following homogenous transformation

$$\begin{bmatrix} \mathbf{b} \\ 1 \end{bmatrix} = {}^0\mathbf{H}_1 {}^1\mathbf{H}_2 {}^2\mathbf{H}_3 \begin{bmatrix} \mathbf{a} \\ 1 \end{bmatrix}. \tag{3.1}$$

3.1 Vector Parameters of a Kinematic Pair

Vector parameters will be used for the geometric description of a robot mechanism. For simplicity we shall limit our consideration to the mechanisms with either parallel or perpendicular consecutive joint axes. Such mechanisms are by far the most frequent in industrial robotics.

In Fig. 3.2, a kinematic pair is shown consisting of two consecutive segments of a robot mechanism, segment $i-1$ and segment i. The two segments are connected by the joint i including both translation and rotation. The relative pose of the joint is determined by the segment vector $\mathbf{b}_{i-1}$ and unit joint vector $\mathbf{e}_i$, as shown in Fig. 3.2. The segment i can be translated with respect to the segment $i-1$ along the vector $\mathbf{e}_i$ for the distance d_i and can be rotated around $\mathbf{e}_i$ for the angle ϑ_i. The coordinate

M. Mihelj et al., *Robotics*, https://doi.org/10.1007/978-3-319-72911-4_3

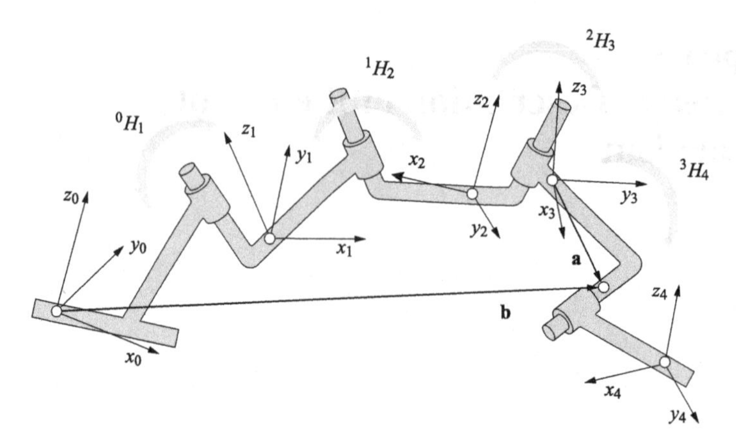

Fig. 3.1 Robot mechanism with coordinate frames attached to its segments

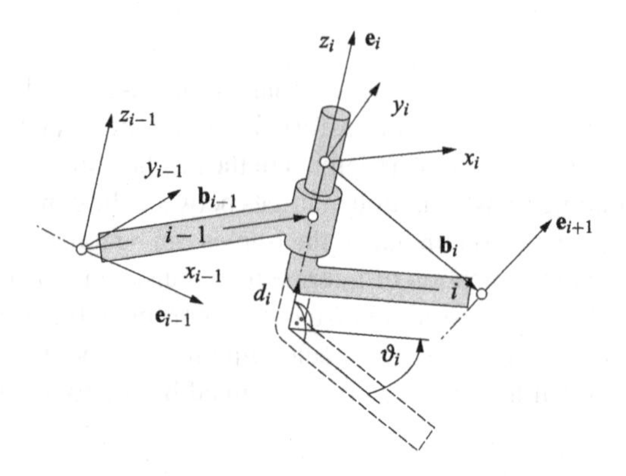

Fig. 3.2 Vector parameters of a kinematic pair

frame x_i–y_i–z_i is attached to the segment i, while the frame x_{i-1}–y_{i-1}–z_{i-1} belongs to the segment $i-1$.

The coordinate frame x_i–y_i–z_i is placed into the axis of the joint i in such a way that it is parallel to the previous frame x_{i-1}–y_{i-1}–z_{i-1} when the kinematic pair is in its initial pose (both joint variables are zero $\vartheta_i = 0$ and $d_i = 0$).

The geometric relations and the relative displacement of two neighboring segments of a robot mechanism are determined by the following parameters:

$\mathbf{e}_i$ —unit vector describing either the axis of rotation or direction of translation in the joint i and is expressed as one of the axes of the frame x_i–y_i–z_i. Its components are the following

$$\mathbf{e}_i = \begin{bmatrix} 1 \\ 0 \\ 0 \end{bmatrix} \text{ or } \begin{bmatrix} 0 \\ 1 \\ 0 \end{bmatrix} \text{ or } \begin{bmatrix} 0 \\ 0 \\ 1 \end{bmatrix};$$

$\mathbf{b}_{i-1}$ —segment vector describing the segment $i-1$ expressed in the frame x_{i-1}–y_{i-1}–z_{i-1}. Its components are the following

$$\mathbf{b}_{i-1} = \begin{bmatrix} b_{i-1,x} \\ b_{i-1,y} \\ b_{i-1,z} \end{bmatrix};$$

ϑ_i —rotational variable representing the angle measured around the $\mathbf{e}_i$ axis in the plane which is perpendicular to $\mathbf{e}_i$ (the angle is zero when the kinematic pair is in the initial position);

d_i —translational variable representing the distance measured along the direction of $\mathbf{e}_i$ (the distance equals zero when the kinematic pair is in the initial position).

If the joint is only rotational (Fig. 3.3 above), the joint variable is represented by the angle ϑ_i, while $d_i = 0$. When the robot mechanism is in its initial pose, the joint angle equals zero $\vartheta_i = 0$ and the coordinate frames x_i–y_i–z_i and x_{i-1}–y_{i-1}–z_{i-1} are parallel. If the joint is only translational (Fig. 3.3 below), the joint variable is d_i,

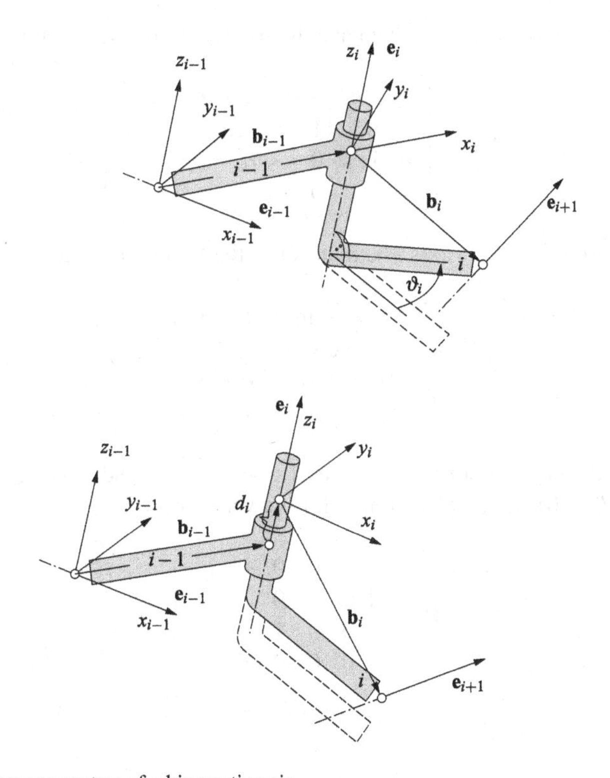

Fig. 3.3 Vector parameters of a kinematic pair

while $\vartheta_i = 0$. When the joint is in its initial position, then $d_i = 0$. In this case the coordinate frames x_i–y_i–z_i and x_{i-1}–y_{i-1}–z_{i-1} are parallel irrespective of the value of the translational variable d_i.

By changing the value of the rotational joint variable ϑ_i, the coordinate frame x_i–y_i–z_i is rotated together with the segment i with respect to the preceding segment $i-1$ and the corresponding frame x_{i-1}–y_{i-1}–z_{i-1}. By changing the translational variable d_i, the displacement is translational, where only the distance between the two neighboring frames is changing.

The transformation between the coordinate frames x_{i-1}–y_{i-1}–z_{i-1} and x_i–y_i–z_i is determined by the homogenous transformation matrix taking one of the three possible forms regarding the direction of the joint vector $\mathbf{e}_i$. When the unit vector $\mathbf{e}_i$ is parallel to the x_i axis, there is

$$
{}^{i-1}\mathbf{H}_i = \begin{bmatrix} 1 & 0 & 0 & d_i + b_{i-1,x} \\ 0 & \cos\vartheta_i & -\sin\vartheta_i & b_{i-1,y} \\ 0 & \sin\vartheta_i & \cos\vartheta_i & b_{i-1,z} \\ 0 & 0 & 0 & 1 \end{bmatrix}, \tag{3.2}
$$

when $\mathbf{e}_i$ is parallel to the y_i axis, we have the following transformation matrix

$$
{}^{i-1}\mathbf{H}_i = \begin{bmatrix} \cos\vartheta_i & 0 & \sin\vartheta_i & b_{i-1,x} \\ 0 & 1 & 0 & d_i + b_{i-1,y} \\ -\sin\vartheta_i & 0 & \cos\vartheta_i & b_{i-1,z} \\ 0 & 0 & 0 & 1 \end{bmatrix} \tag{3.3}
$$

When $\mathbf{e}_i$ is parallel to the z_i axis, the matrix has the following form

$$
{}^{i-1}\mathbf{H}_i = \begin{bmatrix} \cos\vartheta_i & -\sin\vartheta_i & 0 & b_{i-1,x} \\ \sin\vartheta_i & \cos\vartheta_i & 0 & b_{i-1,y} \\ 0 & 0 & 1 & d_i + b_{i-1,z} \\ 0 & 0 & 0 & 1 \end{bmatrix}. \tag{3.4}
$$

In the initial pose the coordinate frames x_{i-1}–y_{i-1}–z_{i-1} and x_i–y_i–z_i are parallel ($\vartheta_i = 0$ and $d_i = 0$) and displaced only for the vector $\mathbf{b}_{i-1}$

$$
{}^{i-1}\mathbf{H}_i = \begin{bmatrix} 1 & 0 & 0 & b_{i-1,x} \\ 0 & 1 & 0 & b_{i-1,y} \\ 0 & 0 & 1 & b_{i-1,z} \\ 0 & 0 & 0 & 1 \end{bmatrix}. \tag{3.5}
$$

3.2 Vector Parameters of the Mechanism

The vector parameters of a robot mechanism are determined in the following four steps:

step 1 —the robot mechanism is placed into the desired initial (reference) pose. The joint axes must be parallel to one of the axes of the reference coordinate frame x_0–y_0–z_0 attached to the robot base. In the reference pose all values of joint variables equal zero, $\vartheta_i = 0$ and $d_i = 0$, $i = 1, 2, \ldots, n$;

step 2 —the centers of the joints $i = 1, 2, \ldots, n$ are selected. The center of joint i can be anywhere along the corresponding joint axis. A local coordinate frame x_i–y_i–z_i is placed into the joint center in such a way that its axes are parallel to the axes of the reference frame x_0–y_0–z_0. The local coordinate frame x_i–y_i–z_i is displaced together with the segment i;

step 3 —the unit joint vector $\mathbf{e}_i$ is allocated to each joint axis $i = 1, 2, \ldots, n$. It is directed along one of the axes of the coordinate frame x_i–y_i–z_i. In the direction of this vector the translational variable d_i is measured, while the rotational variable ϑ_i is assessed around the joint vector $\mathbf{e}_i$;

step 4 —the segment vectors $\mathbf{b}_{i-1}$ are drawn between the origins of the frames x_i–y_i–z_i, $i = 1, 2, \ldots, n$. The segment vector $\mathbf{b}_n$ connects the origin of the frame x_n–y_n–z_n with the robot end-point.

Sometimes an additional coordinate frame is positioned in the reference point of a gripper and denoted as x_{n+1}–y_{n+1}–z_{n+1}. There exists no degree of freedom between the frames x_n–y_n–z_n and x_{n+1}–y_{n+1}–z_{n+1}, as both frames are attached to the same segment. The transformation between them is therefore constant.

The approach to geometric modeling of robot mechanisms will be illustrated by an example of a robot mechanism with four degrees of freedom shown in Fig. 3.4. The selected initial pose of the mechanism together with the marked positions of the joint centers is presented in Fig. 3.5. The corresponding vector parameters and joint variables are gathered in Table 3.1.

The rotational variables ϑ_1, ϑ_2 and ϑ_4 are measured in the planes perpendicular to the joint axes $\mathbf{e}_1$, $\mathbf{e}_2$ and $\mathbf{e}_4$, while the translational variable d_i is measured along the axis $\mathbf{e}_3$. Their values are zero when the robot mechanism is in its initial pose. In Fig. 3.6 the robot manipulator is shown in a pose where all four variables are positive and nonzero. The variable ϑ_1 represents the angle between the initial and momentary y_1 axis, the variable ϑ_2 the angle between the initial and momentary z_2 axis, variable d_3 is the distance between the initial and actual position of the x_3 axis, while ϑ_4 represents the angle between the initial and momentary x_4 axis.

The selected vector parameters of the robot mechanism are inserted into the homogenous transformation matrices (3.2)–(3.4)

$$
{}^0\mathbf{H}_1 = \begin{bmatrix} c1 & -s1 & 0 & 0 \\ s1 & c1 & 0 & 0 \\ 0 & 0 & 1 & h_0 \\ 0 & 0 & 0 & 1 \end{bmatrix},
$$

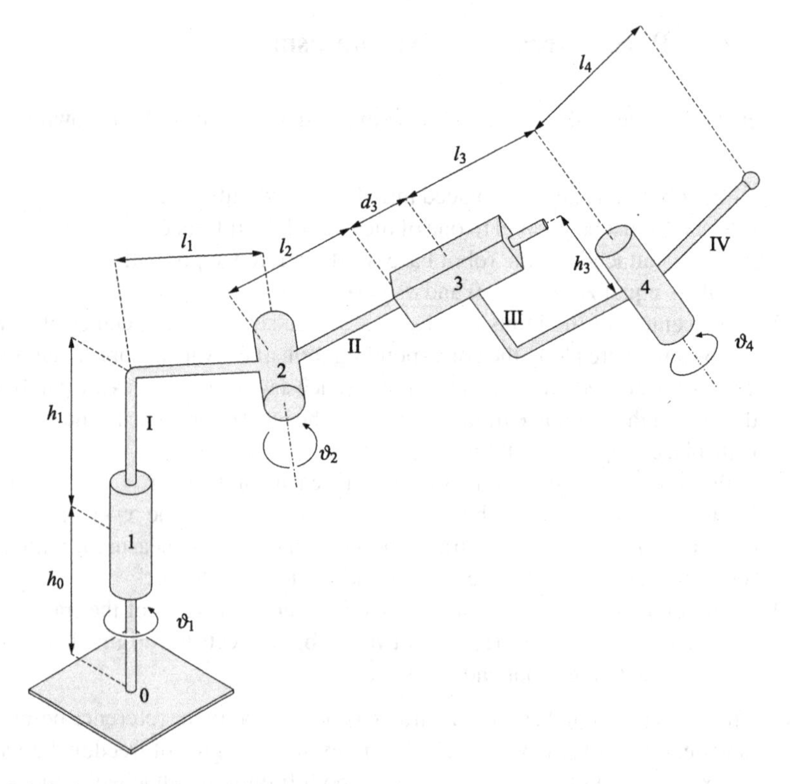

Fig. 3.4 Robot mechanism with four degrees of freedom

$$
{}^1\mathbf{H}_2 = \begin{bmatrix} 1 & 0 & 0 & 0 \\ 0 & c2 & -s2 & l_1 \\ 0 & s2 & c2 & h_1 \\ 0 & 0 & 0 & 1 \end{bmatrix},
$$

$$
{}^2\mathbf{H}_3 = \begin{bmatrix} 1 & 0 & 0 & 0 \\ 0 & 1 & 0 & d_3 + l_2 \\ 0 & 0 & 1 & 0 \\ 0 & 0 & 0 & 1 \end{bmatrix},
$$

$$
{}^3\mathbf{H}_4 = \begin{bmatrix} c4 & -s4 & 0 & 0 \\ s4 & c4 & 0 & l_3 \\ 0 & 0 & 1 & -h_3 \\ 0 & 0 & 0 & 1 \end{bmatrix}.
$$

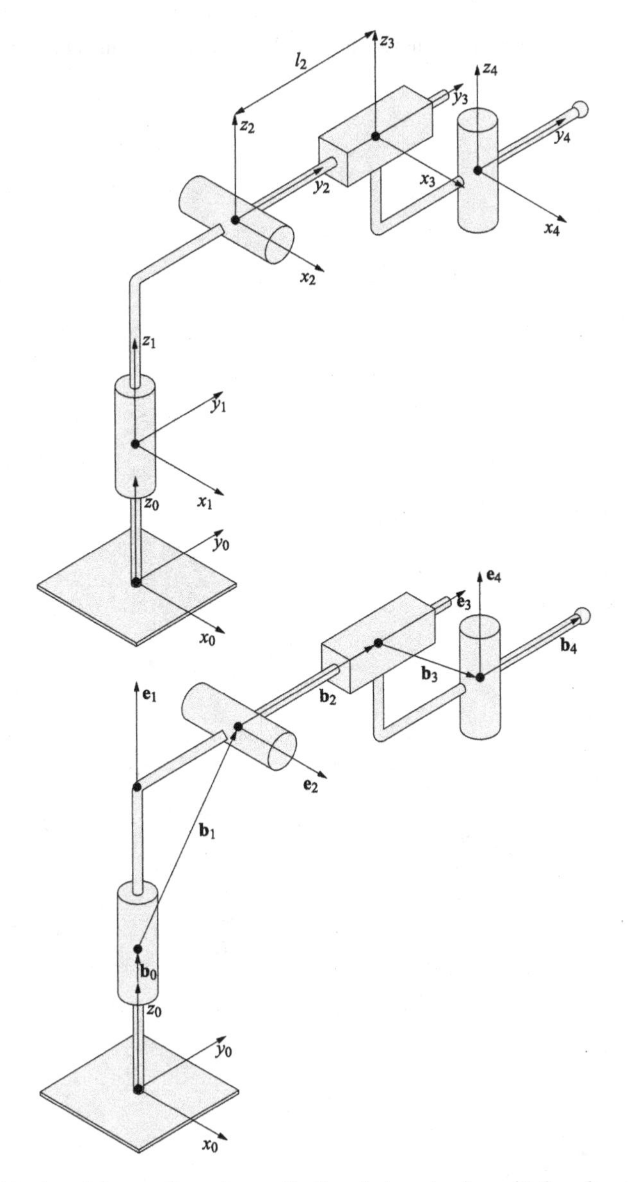

Fig. 3.5 Positioning of the coordinate frames for the robot mechanism with four degrees of freedom

An additional homogenous matrix describes the position of the gripper reference point where the coordinate frame x_5–y_5–z_5 can be allocated

$$ {}^4\mathbf{H}_5 = \begin{bmatrix} 1 & 0 & 0 & 0 \\ 0 & 1 & 0 & l_4 \\ 0 & 0 & 1 & 0 \\ 0 & 0 & 0 & 1 \end{bmatrix}. $$

Table 3.1 Vector parameters and joint variables for the robot mechanism in Fig. 3.5

i	1	2	3	4
ϑ_i	ϑ_1	ϑ_2	0	ϑ_4
d_i	0	0	d_3	0

i	1	2	3	4
	0	1	0	0
$\mathbf{e}_i$	0	0	1	0
	1	0	0	1

i	1	2	3	4	5
	0	0	0	0	0
$\mathbf{b}_{i-1}$	0	l_1	l_2	l_3	l_4
	h_0	h_1	0	$-h_3$	0

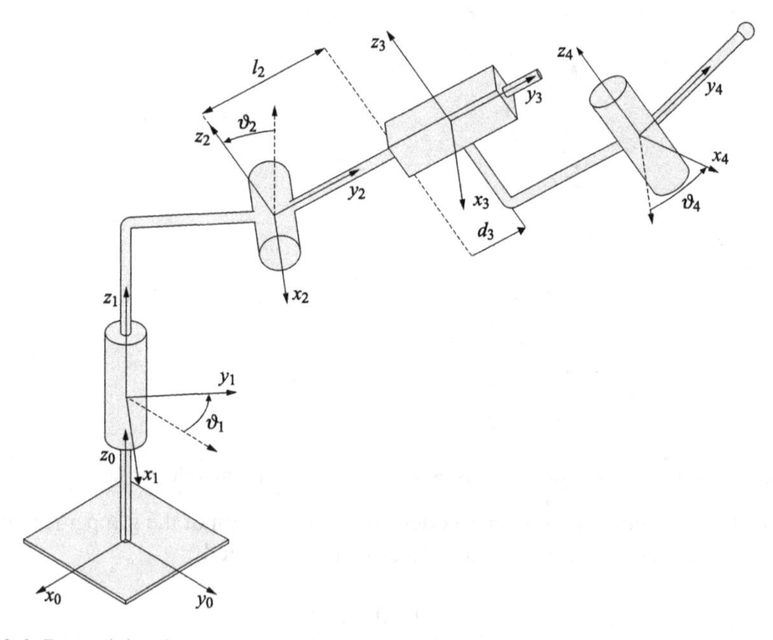

Fig. 3.6 Determining the rotational and translational variables for the robot mechanism with four degrees of freedom

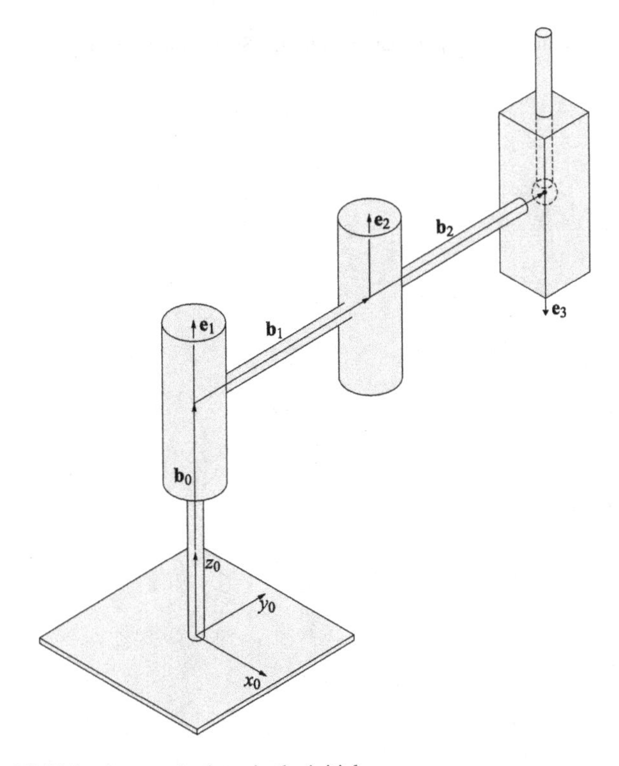

Fig. 3.7 The SCARA robot manipulator in the initial pose

This last matrix is constant as the frames x_4–y_4–z_4 and x_5–y_5–z_5 are parallel and displaced for the distance l_4. Usually this additional frame is not even attached to the robot mechanism, as the position and orientation of the gripper can be described in the frame x_4–y_4–z_4.

When determining the initial (home) pose of the robot mechanism we must take care that the joint axes are parallel to one of the axes of the reference coordinate frame. The initial pose should be selected in such a way that it is simple and easy to examine, that it corresponds well to the anticipated robot tasks and that it minimizes the number of required mathematical operations included in the transformation matrices.

As another example we shall consider the SCARA robot manipulator whose geometric model was developed already in the previous chapter and is shown in Fig. 2.10. The robot mechanism should be first positioned into the initial pose in such a way that the joint axes are parallel to one of the axes of the reference frame x_0–y_0–z_0. In this way the two neighboring segments are either parallel or perpendicular. The translational joint must be in its initial position ($d3 = 0$). The SCARA robot in the selected initial pose is shown in Fig. 3.7.

The joint coordinate frames x_i–y_i–z_i are all parallel to the reference frame. Therefore, we shall draw only the reference frame and have the dots indicate the joint centers. In the centers of both rotational joints, unit vectors $\mathbf{e}_1$ and $\mathbf{e}_2$ are placed

Table 3.2 Vector parameters and joint variables for the SCARA robot manipulator

i	1	2	3	4
ϑ_i	ϑ_1	ϑ_2	0	ϑ_4
d_i	0	0	d_3	0

i	1	2	3	4
	0	1	0	0
$\mathbf{e}_i$	0	0	1	0
	1	0	0	1

i	1	2	3	4	5
	0	0	0	0	0
$\mathbf{b}_{i-1}$	0	l_1	l_2	l_3	l_4
	h_0	h_1	0	$-h_3$	0

along the joint axes. The rotation around the $\mathbf{e_1}$ vector is described by the variable ϑ_1, while ϑ_2 represents the angle about the $\mathbf{e_2}$ vector. Vector $\mathbf{e_3}$ is placed along the translational axis of the third joint. Its translation variable is described by d_3. The first joint is connected to the robot base by the vector $\mathbf{b_0}$. Vector $\mathbf{b_1}$ connects the first and the second joint and vector $\mathbf{b_2}$ the second and the third joint. The variables and vectors are gathered in the three tables (Table 3.2).

In our case all $\mathbf{e_i}$ vectors are parallel to the z_0 axis, the homogenous transformation matrices are therefore written according Eq. (3.4). Similar matrices are obtained for both rotational joints.

$$^0\mathbf{H}_1 = \begin{bmatrix} c1 & -s1 & 0 & 0 \\ s1 & c1 & 0 & 0 \\ 0 & 0 & 1 & l_1 \\ 0 & 0 & 0 & 1 \end{bmatrix}.$$

$$^1\mathbf{H}_2 = \begin{bmatrix} c2 & -s2 & 0 & 0 \\ s2 & c2 & 0 & l_2 \\ 0 & 0 & 1 & 0 \\ 0 & 0 & 0 & 1 \end{bmatrix}.$$

For the translational joint, $\vartheta_3 = 0$ must be inserted into Eq. (3.4), giving

$$
{}^{2}\mathbf{H}_3 = \begin{bmatrix} 1 & 0 & 0 & 0 \\ 0 & 1 & 0 & l_3 \\ 0 & 0 & 1 & -d_3 \\ 0 & 0 & 0 & 1 \end{bmatrix}.
$$

With postmultiplication of all three matrices the geometric model of the SCARA robot is obtained

$$
{}^{0}\mathbf{H}_3 = {}^{0}\mathbf{H}_1\,{}^{1}\mathbf{H}_2\,{}^{2}\mathbf{H}_3 = \begin{bmatrix} c12 & -s12 & 0 & -l_3 s12 - l_2 s1 \\ s12 & c12 & 0 & l_3 c12 + l_2 c1 \\ 0 & 0 & 1 & l_1 - d_3 \\ 0 & 0 & 0 & 1 \end{bmatrix}.
$$

We obtained the same result as in previous chapter, however in a much simpler and more clearer way.

Chapter 4
Orientation

We often describe our environment as a three-dimensional world. The world of the roboticist is, however, six-dimensional. He must not only consider the position of an object, but also its orientation. When a robot gripper or end-effector approaches an object to be grasped, the space angles between the gripper and the object are of the utmost importance.

Six parameters are required to completely describe the position and orientation of an object in a space. Three parameters refer to the position and the other three to the orientation of the object. There are three possible ways how to mathematically describe the orientation of the object. The first possibility is a rotation/orientation matrix consisting of nine elements. The matrix represents a redundant description of the orientation. A non-redundant description is given by RPY or Euler angles. In both cases we have three angles. The RPY angles are defined about the axes of a fixed coordinate frame, while the Euler angles describe the orientation about a relative coordinate frame. The third possible description of the orientation is enabled by four parameters of quaternion.

In the second chapter we already became acquainted with rotation matrices around x, y, and z axis of a rectangular frame. We found them useful when developing the geometrical model of a robot mechanism. It is not difficult to understand that there exists also a matrix describing the rotation around an arbitrary axis. This can be expressed in the following form

$$
{}^{0}\mathbf{R}_{1} = \begin{bmatrix} {}^{1}\mathbf{i}^{0}\mathbf{i} & {}^{1}\mathbf{j}^{0}\mathbf{i} & {}^{1}\mathbf{k}^{0}\mathbf{i} \\ {}^{1}\mathbf{i}^{0}\mathbf{j} & {}^{1}\mathbf{j}^{0}\mathbf{j} & {}^{1}\mathbf{k}^{0}\mathbf{j} \\ {}^{1}\mathbf{i}^{0}\mathbf{k} & {}^{1}\mathbf{j}^{0}\mathbf{k} & {}^{1}\mathbf{k}^{0}\mathbf{k} \end{bmatrix}. \tag{4.1}
$$

The matrix of the dimension 3×3 does not only represent the rotation, but also the orientation of the frame x_1–y_1–z_1 with respect to the frame x_0–y_0–z_0, as it can be seen from Fig. 4.1. The reference frame x_0–y_0–z_0 is described by the unit vectors

M. Mihelj et al., *Robotics*, https://doi.org/10.1007/978-3-319-72911-4_4

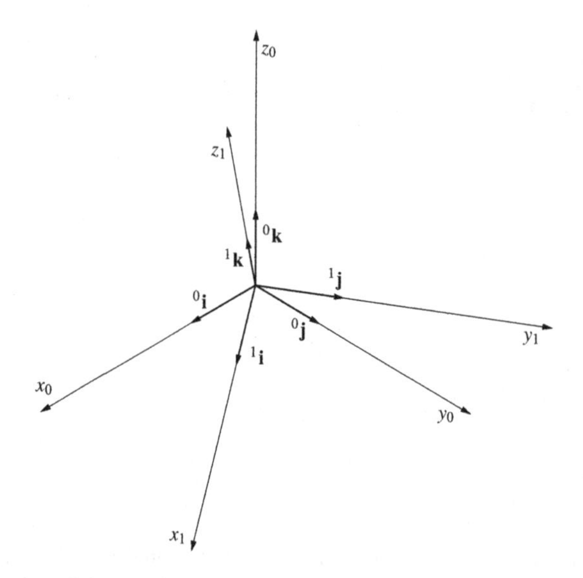

Fig. 4.1 Orientation of the coordinate frame x_1–y_1–z_1 with respect to the reference coordinate frame x_0–y_0–z_0

${}^0\mathbf{i}$, ${}^0\mathbf{j}$, and ${}^0\mathbf{k}$ and the rotated frame x_1–y_1–z_1 with the unit vectors ${}^1\mathbf{i}$, ${}^1\mathbf{j}$, and ${}^1\mathbf{k}$. Both coordinate frames coincide in the same origin. As we are dealing with the unit vectors, the elements of the rotation/orientation matrix are simply the cosines of the angles appertaining to each pair of the axes.

Let us consider the example from Fig. 4.2 and calculate the matrix representing the orientation of the frame x_1–y_1–z_1, which is rotated for the angle $+\vartheta$ with respect to the frame x_0–y_0–z_0.

We are dealing with the following non-zero products of the unit vectors

$$\begin{aligned} {}^0\mathbf{i}\,{}^1\mathbf{i} &= 1, \\ {}^0\mathbf{j}\,{}^1\mathbf{j} &= \cos\vartheta, \\ {}^0\mathbf{k}\,{}^1\mathbf{k} &= \cos\vartheta, \\ {}^0\mathbf{j}\,{}^1\mathbf{k} &= -\sin\vartheta, \\ {}^0\mathbf{k}\,{}^1\mathbf{j} &= \sin\vartheta. \end{aligned} \tag{4.2}$$

The matrix describing the orientation of the frame x_1–y_1–z_1 with respect to x_0–y_0–z_0 is therefore

$$\mathbf{R}_x = \begin{bmatrix} 1 & 0 & 0 \\ 0 & c\vartheta & -s\vartheta \\ 0 & s\vartheta & c\vartheta \end{bmatrix} \tag{4.3}$$

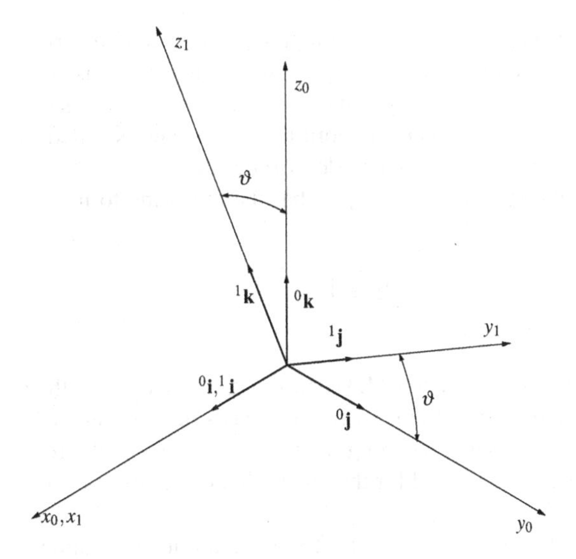

Fig. 4.2 Two coordinate frames rotated about the x_0 axis

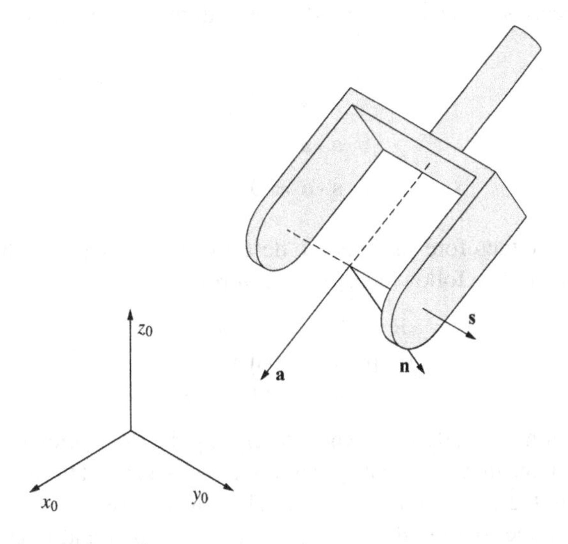

Fig. 4.3 Orientation of robot gripper

The matrix (4.3) can be interpreted also as the rotation matrix around the x axis that we already know as part of the homogeneous matrix (2.6) from the second chapter.

The notion of orientation is in robotics mostly related to the orientation of the robot gripper. A coordinate frame with three unit vectors **n**, **s**, and **a**, describing the orientation of the gripper, is placed between two fingers of a simple robot gripper (Fig. 4.3).

The z axis vector lays in the direction of the approach of the gripper to the object. It is therefore denoted by vector **a** (approach). Vector, which is aligned with y axis, describes the direction of sliding of the fingers and is denoted as **s** (slide). The third vector completes the right-handed coordinate frame and is called normal. This can be shown as $\mathbf{n} = \mathbf{s} \times \mathbf{a}$. The matrix describing the orientation of the gripper with respect to the reference frame x_0–y_0–z_0 has the following form

$$\mathbf{R} = \begin{bmatrix} n_x & s_x & a_x \\ n_y & s_y & a_y \\ n_z & s_z & a_z \end{bmatrix}. \tag{4.4}$$

The element n_x of the matrix (4.3) denotes the projection of the unit vector **n** on the x_0 axis of the reference frame. It equals the cosine of the angle between the axes x and x_0 and has the same meaning as the element $^1\mathbf{i}^0\mathbf{i}$ of the rotation/orientation matrix (4.1). The same is valid for the eight other elements of the orientation matrix **R** (4.3).

To describe the orientation of an object we do not need nine elements of the matrix. The left column vector is the cross product of vectors **s** and **a**. The vectors **s** and **a** are unit vectors which are perpendicular with respect to each other, so that we have

$$\begin{aligned} \mathbf{s} \cdot \mathbf{s} &= 1, \\ \mathbf{a} \cdot \mathbf{a} &= 1, \\ \mathbf{s} \cdot \mathbf{a} &= 0. \end{aligned} \tag{4.5}$$

Three elements are therefore sufficient to describe the orientation. This orientation is often described by the following sequence of rotations

R - roll - about z axis,
P - pitch - about y axis,
Y - yaw - about x axis.

This description is mostly used when describing the orientation of a ship or airplane. Let us imagine that the airplane flies along z axis and that the coordinate frame is positioned into the center of the airplane. Then, R represents the rotation φ about z axis, P refers to the rotation ϑ about y axis and Y to the rotation ψ about x axis, as shown in Fig. 4.4.

The use of the RPY angles for a robot gripper is shown in Fig. 4.5. As it can be realized from Figs. 4.4 and 4.5, the RPY orientation is defined with respect to a fixed coordinate frame. When developing the geometrical model of the SCARA robot manipulator in the second chapter, we were postmultiplying the homogenous transformation matrices describing the rotation (or translation) of each particular joint. The position and orientation of each joint frame was defined with respect to the preceding frame, appertaining to the joint axis which is not fixed. In this case, as

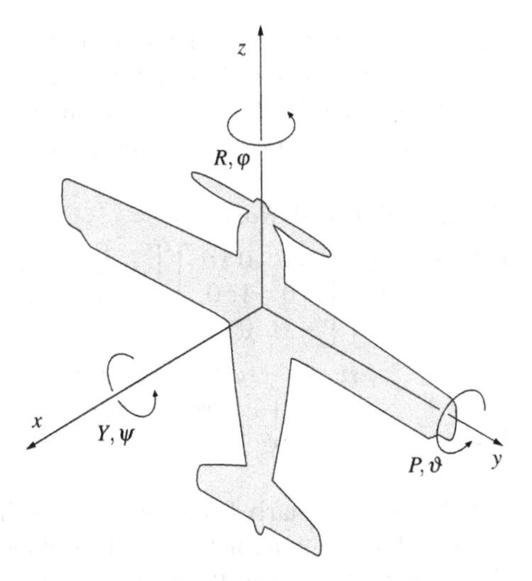

Fig. 4.4 RPY angles for the case of an airplane

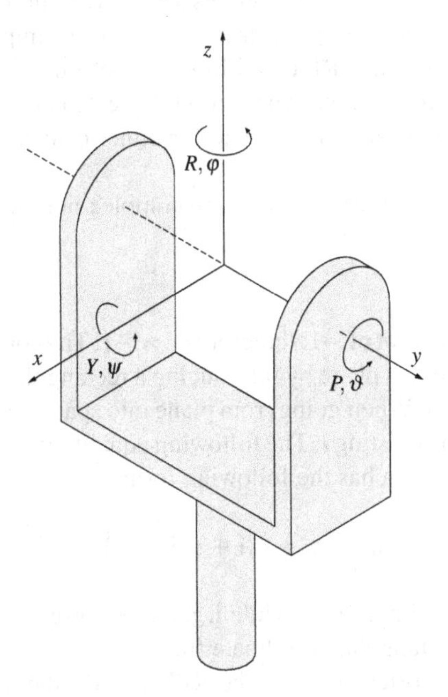

Fig. 4.5 RPY angles for the case of robot gripper

we have seen, we are multiplying the matrices from left to right. When we are dealing with consecutive rotations about the axes of the same coordinate frame, we make use of the premultiplication of the rotation matrices. In other words, the multiplications are performed in the reverse order from right to left.

We start with the rotation φ about z axis, continue with rotation ϑ around y axis and finish with the rotation ψ about x axis. The reverse order of rotations is also evident from the naming of RPY angles. The orientation matrix, which belongs to RPY angles, is obtained by the following multiplication of the rotation matrices

$$\begin{aligned}
\mathbf{R}(\varphi, \vartheta, \psi) &= Rot(z, \varphi)Rot(y, \vartheta)Rot(x, \psi) = \\
&= \begin{bmatrix} c\varphi & -s\varphi & 0 \\ s\varphi & c\varphi & 0 \\ 0 & 0 & 1 \end{bmatrix} \begin{bmatrix} c\vartheta & 0 & s\vartheta \\ 0 & 1 & 0 \\ -s\vartheta & 0 & c\vartheta \end{bmatrix} \begin{bmatrix} 1 & 0 & 0 \\ 0 & c\psi & -s\psi \\ 0 & s\psi & c\psi \end{bmatrix} = \\
&= \begin{bmatrix} c\varphi c\vartheta & c\varphi s\vartheta s\psi - s\varphi c\psi & c\varphi s\vartheta c\psi + s\varphi s\psi \\ s\varphi s\vartheta & s\varphi s\vartheta s\psi + c\varphi c\psi & s\varphi s\vartheta c\psi - c\varphi c\psi \\ -s\vartheta & c\vartheta s\psi & c\vartheta c\psi \end{bmatrix}.
\end{aligned} \tag{4.6}$$

Equation (4.6) calculates the rotation matrix from the corresponding RPY angles.

We learned that rotation and orientation can be described either by rotation matrices or by RPY angles. In the first case we need 9 parameters, while only 3 parameters are required in the latter case. While matrices are convenient for computations, they do not however, provide a fast and clear image of, for example, the orientation of a robot gripper within a space. RPY and Euler angles do nicely present the orientation of a gripper, but they are not appropriate for calculations. In this chapter we shall learn that quaternions are appropriate for either calculation or description of orientation.

The quaternions represent extension of the complex numbers

$$z = a + \mathbf{i}b, \tag{4.7}$$

where $\mathbf{i}$ means the square root of -1, therefore $\mathbf{i}^2 = -1$. The complex numbers can be geometrically presented in a plane by introducing a rectangular frame with $\Re e$ (real) and $\Im m$ (imaginary) axis. When going from plane into space, two unit vectors $\mathbf{j}$ and $\mathbf{k}$ must be added to already existing $\mathbf{i}$. The following equality $\mathbf{i}^2 = \mathbf{j}^2 = \mathbf{k}^2 = \mathbf{ijk} = -1$ is also valid. The quaternion has the following form

$$q = q_0 + q_1\mathbf{i} + q_2\mathbf{j} + q_3\mathbf{k}. \tag{4.8}$$

In the Eq. (4.8) q_i are real numbers, while $\mathbf{i}$, $\mathbf{j}$, and $\mathbf{k}$ correspond to the unit vectors along the axes of the rectangular coordinate frame.

When describing the orientation by the RPY angles, the multiplications of the rotation matrices were needed. In a similar way we need to multiply the quaternions

$$pq = (p_0 + p_1\mathbf{i} + p_2\mathbf{j} + p_3\mathbf{k})(q_0 + q_1\mathbf{i} + q_2\mathbf{j} + q_3\mathbf{k}). \tag{4.9}$$

Table 4.1 Rules for quaternion multiplications

*	1	**i**	**j**	**k**
1	1	**i**	**j**	**k**
i	**i**	−1	**k**	−**j**
j	**j**	−**k**	−1	**i**
k	**k**	**j**	−**i**	−1

The multiplication of quaternions is not commutative. When multiplying two quaternions we shall make use of the Table 4.1. Let us multiply two quaternions

$$\begin{aligned}
&(2+3\mathbf{i}-\mathbf{j}+5\mathbf{k})(3-4\mathbf{i}+2\mathbf{j}+\mathbf{k}) = \\
&= 6+9\mathbf{i}-3\mathbf{j}+15\mathbf{k}- \\
&-8\mathbf{i}-12\mathbf{i}^2+4\mathbf{ji}-20\mathbf{ki}+ \\
&+4\mathbf{j}+6\mathbf{ij}-2\mathbf{j}^2+10\mathbf{kj}+ \\
&+2\mathbf{k}+3\mathbf{ik}-\mathbf{jk}+5\mathbf{k}^2 = \\
&= 6+9\mathbf{i}-3\mathbf{j}+15\mathbf{k}- \\
&-8\mathbf{i}+12-4\mathbf{k}-20\mathbf{j}+ \\
&+4\mathbf{j}+6\mathbf{k}+2-10\mathbf{i}+ \\
&+2\mathbf{k}-3\mathbf{j}-\mathbf{i}-5 = \\
&= 15-10\mathbf{i}-22\mathbf{j}+19\mathbf{k}.
\end{aligned} \tag{4.10}$$

The following expression of a quaternion is specially appropriate to describe the orientation in the space

$$q = \cos\frac{\vartheta}{2} + \sin\frac{\vartheta}{2}\mathbf{s}. \tag{4.11}$$

In the Eq. (4.11) **s** is a unit vector aligned with the rotation axis, while ϑ is the angle of rotation. The orientation quaternion can be obtained from the RPY angles. Rotation R is described by the quaternion

$$q_{z\varphi} = \cos\frac{\varphi}{2} + \sin\frac{\varphi}{2}\mathbf{k}. \tag{4.12}$$

The following quaternion belongs to the rotation P

$$q_{y\vartheta} = \cos\frac{\vartheta}{2} + \sin\frac{\vartheta}{2}\mathbf{j}, \tag{4.13}$$

while rotation Y can be written as follows

$$q_{x\psi} = \cos\frac{\psi}{2} + \sin\frac{\psi}{2}\mathbf{i}. \tag{4.14}$$

After multiplying the above three quaternions (4.12–4.14), the resulting orientation quaternion is obtained

$$q(\varphi, \vartheta, \psi) = q_{z\varphi} q_{y\vartheta} q_{x\psi}. \tag{4.15}$$

Let us illustrate the three descriptions of the orientation, i.e. RPY angles, rotation matrix, and quaternions, by an example of description of gripper orientation. To make the example clear and simple, the plane of the two-finger gripper will be placed into the x_0–y_0 plane of the reference frame (Fig. 4.6). The RPY angles can be read from the Fig. 4.6. The rotations around z and y axis equal zero. The rotation for $-60°$ around the x axis can be seen from the Fig. 4.6. The orientation of the gripper can be, therefore, described by the following set of RPY angles

$$\varphi = 0,\ \vartheta = 0,\ \psi = -60°. \tag{4.16}$$

From the Fig. 4.6 we can read also the angles between the axes of the reference and gripper coordinate frame. Their cosines represent the orientation/rotation matrix $\mathbf{R}$

$$\begin{aligned} n_x &= \cos 0°,\ s_x = \cos 90°,\ a_x = \cos 90°, \\ n_y &= \cos 90°,\ s_y = \cos 60°,\ a_y = \cos 30°, \\ n_z &= \cos 0°,\ s_z = \cos 150°,\ a_z = \cos 60°. \end{aligned} \tag{4.17}$$

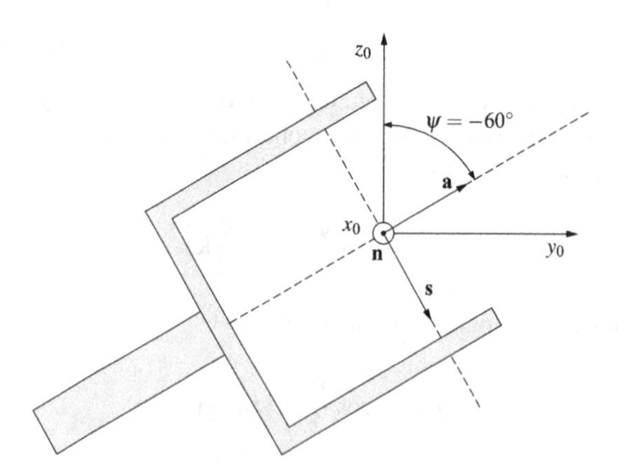

Fig. 4.6 Orientation of robot gripper

The matrix $\mathbf{R}$ can be calculated also by inserting the known RPY angles into the Eq. (4.6)

$$\mathbf{R} = \begin{bmatrix} 1 & 0 & 0 \\ 0 & 0.5 & 0.866 \\ 0 & -0.866 & 0.5 \end{bmatrix}. \tag{4.18}$$

In this way the correctness of our reading of the angles from the Fig. 4.6 was tested. We shall calculate the orientation quaternion by inserting the RPY angles into the Eqs. (4.12–4.14)

$$\begin{aligned} q_{z\varphi} &= 1 + 0\mathbf{k}, \\ q_{y\vartheta} &= 1 + 0\mathbf{j}, \\ q_{x\psi} &= 0.866 - 0.5\mathbf{i}. \end{aligned} \tag{4.19}$$

The orientation quaternion is obtained after multiplying the three above quaternions (4.15)

$$q_0 = 0.866, q_1 = -0.5, q_2 = 0, q_3 = 0. \tag{4.20}$$

The Eqs. (4.16), (4.18) and (4.20) demonstrate three different descriptions of the same gripper orientation.

Chapter 5
Two-Segment Robot Manipulator

5.1 Kinematics

Kinematics is part of classic mechanics that study motion without considering the forces which are responsible for this motion. Motion is in general described by trajectories, velocities and accelerations. In robotics we are mainly interested in trajectories and velocities, as both can be measured by the joint sensors. In robot joints, the trajectories are measured either as the angle in a rotational joint or as the distance in a translational joint. The joint variables are also called internal coordinates. When planning and programming a robot task the trajectory of the robot end-point is of utmost importance. The position and orientation of the end-effector are described by external coordinates. Computation of external variables from the internal variables, and vice versa, is the central problem of robot kinematics.

In this chapter we shall limit our interest to a planar two-segment robot manipulator with two rotational joints (Fig. 5.1). According to the definition given in the introductory chapter, such a mechanism can hardly be called a robot. Nevertheless, this mechanism is an important part of the SCARA and anthropomorphic robot structures and will allow us to study several characteristic properties of the motion of robot mechanisms.

There is a distinction between direct and inverse kinematics. Direct kinematics in the case of a two-segment robot represents the calculation of the position of the robot end-point from the known joint angles. Inverse kinematics calculates the joint variables from the known position of the robot end-point. Direct kinematics represents the simpler problem, as we have a single solution for the position of the robot end-point. The solutions of inverse kinematics depend largely on the structure of the robot manipulator. We often deal with several solutions for the joint variables resulting in the same position of the robot end-point, while in some cases an analytic solution of inverse kinematics does not exist.

Kinematic analysis includes also the relations between the velocity of the robot end-point and the velocities of individual joints. We shall find that inverse kinematics for velocities is simpler than inverse kinematics for trajectories. We shall first find the

M. Mihelj et al., *Robotics*, https://doi.org/10.1007/978-3-319-72911-4_5

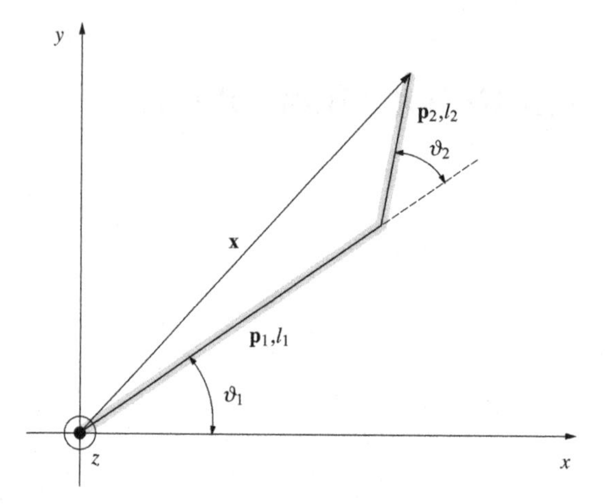

Fig. 5.1 Planar two-segment robot manipulator

solution of direct kinematics for the trajectories. By differentiation we then obtain the equations describing direct kinematics for the velocities. By simple matrix inversion the inverse kinematics for velocities can be computed. Let us now consider the planar two-segment robot manipulator shown in Fig. 5.1.

The axis of rotation of the first joint is presented by the vertical z axis pointing out of the plane. Vector $\mathbf{p}_1$ is directed along the first segment

$$\mathbf{p}_1 = l_1 \begin{bmatrix} \cos\vartheta_1 \\ \sin\vartheta_1 \end{bmatrix}. \tag{5.1}$$

Vector $\mathbf{p}_2$ is along with the second segment. Its components can be read from Fig. 5.1

$$\mathbf{p}_2 = l_2 \begin{bmatrix} \cos(\vartheta_1 + \vartheta_2) \\ \sin(\vartheta_1 + \vartheta_2) \end{bmatrix}. \tag{5.2}$$

Vector $\mathbf{x}$ connects the origin of the coordinate frame with the robot end-point

$$\mathbf{x} = \mathbf{p}_1 + \mathbf{p}_2. \tag{5.3}$$

So we have for the position of the robot end-point

$$\mathbf{x} = \begin{bmatrix} x \\ y \end{bmatrix} = \begin{bmatrix} l_1\cos\vartheta_1 + l_2\cos(\vartheta_1 + \vartheta_2) \\ l_1\sin\vartheta_1 + l_2\sin(\vartheta_1 + \vartheta_2) \end{bmatrix}. \tag{5.4}$$

By defining the vector of joint angles

$$\mathbf{q} = \begin{bmatrix} \vartheta_1 & \vartheta_2 \end{bmatrix}^T, \tag{5.5}$$

the Eq. (5.4) can be written in the following shorter form

$$\mathbf{x} = \mathbf{k}(\mathbf{q}), \tag{5.6}$$

where $\mathbf{k}(\cdot)$ represents the equations of direct kinematics.

The relation between the velocities of the robot end-point and joint velocities is obtained by differentiation. The coordinates of the end-point are functions of the joint angles, which in turn are functions of time

$$\begin{aligned} x &= x(\vartheta_1(t), \vartheta_2(t)) \\ y &= y(\vartheta_1(t), \vartheta_2(t)). \end{aligned} \tag{5.7}$$

By calculating the time derivatives of Eq. (5.7) and arranging them into matrix form, we can write

$$\begin{bmatrix} \dot{x} \\ \dot{y} \end{bmatrix} = \begin{bmatrix} \frac{\partial x}{\partial \vartheta_1} & \frac{\partial x}{\partial \vartheta_2} \\ \frac{\partial y}{\partial \vartheta_1} & \frac{\partial y}{\partial \vartheta_2} \end{bmatrix} \begin{bmatrix} \dot{\vartheta}_1 \\ \dot{\vartheta}_2 \end{bmatrix}. \tag{5.8}$$

For our two-segment robot manipulator we obtain the following expression

$$\begin{bmatrix} \dot{x} \\ \dot{y} \end{bmatrix} = \begin{bmatrix} -l_1 s1 - l_2 s12 & -l_2 s12 \\ l_1 c1 + l_2 c12 & l_2 c12 \end{bmatrix} \begin{bmatrix} \dot{\vartheta}_1 \\ \dot{\vartheta}_2 \end{bmatrix}. \tag{5.9}$$

The matrix, which is in our case of the second order, is called the Jacobian matrix $\mathbf{J}(\mathbf{q})$. The relation (5.9) can be written in short form as

$$\dot{\mathbf{x}} = \mathbf{J}(\mathbf{q})\dot{\mathbf{q}}. \tag{5.10}$$

In this way the problems of direct kinematics for trajectories and velocities are solved. When solving the inverse kinematics, we calculate the joint angles from the known position of the robot end-point. Figure 5.2 shows only those parameters of the two-segment robot mechanism which are relevant for the calculation of the ϑ_2 angle. The cosine rule is used

$$x^2 + y^2 = l_1^2 + l_2^2 - 2l_1 l_2 \cos(180° - \vartheta_2), \tag{5.11}$$

where $-\cos(180° - \vartheta_2) = \cos(\vartheta_2)$. The angle of the second segment of the two-segment manipulator is calculated as the inverse trigonometric function

$$\vartheta_2 = \arccos \frac{x^2 + y^2 - l_1^2 - l_2^2}{2l_1 l_2}. \tag{5.12}$$

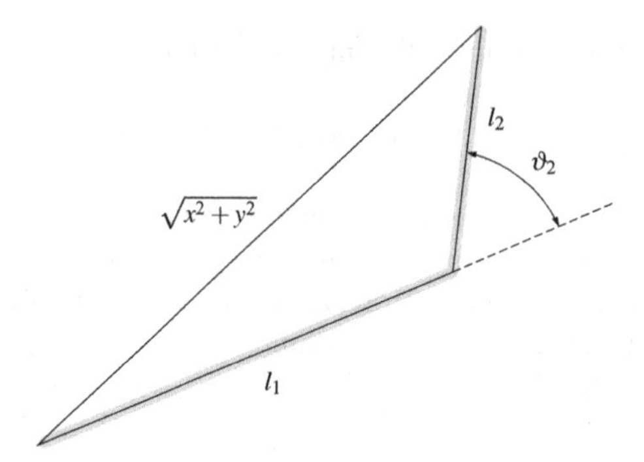

Fig. 5.2 Calculation of the ϑ_2 angle

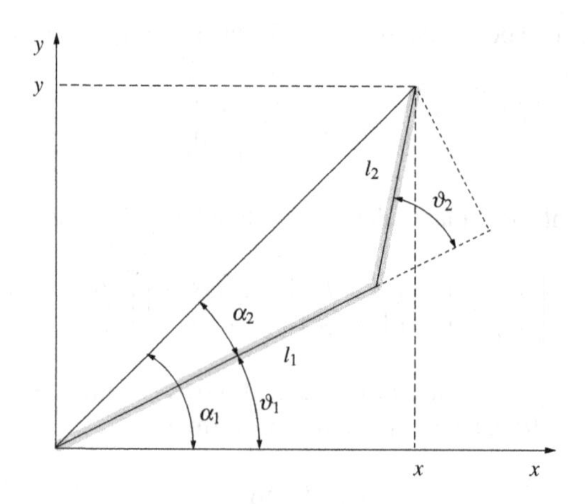

Fig. 5.3 Calculation of the ϑ_1 angle

The angle of the first segment is calculated with the aid of Fig. 5.3. It is obtained as the difference of angles α_1 and α_2

$$\vartheta_1 = \alpha_1 - \alpha_2.$$

The angle α_1 is obtained from the right-angle triangle made of horizontal x and vertical y coordinates of the robot end-point. The angle α_2 is obtained by elongating the triangle of Fig. 5.2 into the right-angle triangle, as shown in Fig. 5.3. Again we make use of the inverse trigonometric functions

$$\vartheta_1 = \arctan\left(\frac{y}{x}\right) - \arctan\left(\frac{l_2 \sin \vartheta_2}{l_1 + l_2 \cos \vartheta_2}\right). \tag{5.13}$$

When calculating the ϑ_2 angle, we have two solutions, *elbow-up* and *elbow-down*, as shown in Fig. 5.4. A degenerate solution is represented by the end-point

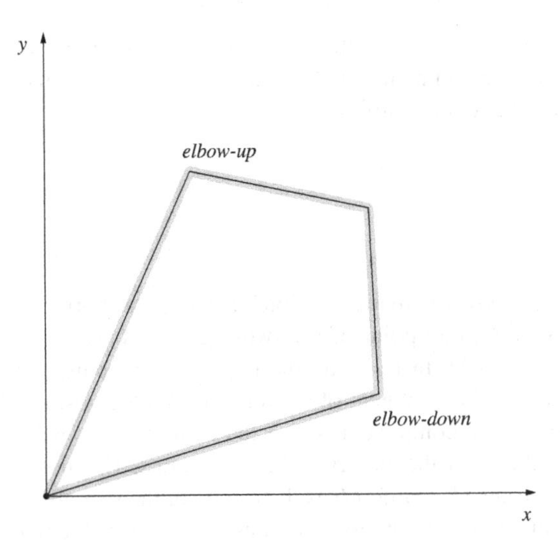

Fig. 5.4 Two solutions of inverse kinematics

position $x = y = 0$ when both segments are of equal length $l_1 = l_2$. In this case $\arctan\left(\frac{y}{x}\right)$ is not defined. When the angle $\vartheta_2 = 180°$, the base of the simple two-segment mechanism can be reached at an arbitrary angle ϑ_1. However, when a point (x, y) lies outside of the manipulator workspace, the problem of inverse kinematics cannot be solved.

The relation between the joint velocities and the velocity of the end-point is obtained by inverting the Jacobian matrix $\mathbf{J}(\mathbf{q})$

$$\dot{\mathbf{q}} = \mathbf{J}^{-1}(\mathbf{q})\dot{\mathbf{x}}. \tag{5.14}$$

The matrices of order 2×2 can be inverted as follows

$$\mathbf{A} = \begin{bmatrix} a & b \\ c & d \end{bmatrix} \quad \mathbf{A}^{-1} = \frac{1}{ad - cb}\begin{bmatrix} d & -b \\ -c & a \end{bmatrix}.$$

For our two-segment manipulator we can write

$$\begin{bmatrix} \dot{\vartheta}_1 \\ \dot{\vartheta}_2 \end{bmatrix} = \frac{1}{l_1 l_2\, s2}\begin{bmatrix} l_2\, c12 & l_2\, s12 \\ -l_1\, c1 - l_2\, c12 & -l_1\, s1 - l_2\, s12 \end{bmatrix}\begin{bmatrix} \dot{x} \\ \dot{y} \end{bmatrix}. \tag{5.15}$$

In general examples of robot manipulators, it is not necessary that the Jacobian matrix has the quadratic form. In this case, the so called pseudoinverse matrix $(\mathbf{J}\mathbf{J}^T)^{-1}$ is calculated. For a robot with six degrees of freedom the Jacobian matrix is quadratic, however after inverting, it becomes rather impractical. When the manipulator is close to singular poses (e.g., when the angle ϑ_2 is close to zero for the simple two-segment robot), the inverse Jacobian matrix is ill defined. We shall make use of the Jacobian matrix when studying robot control.

5.2 Statics

After the end of the robot kinematics section let us make a short leap to robot statics. Let us suppose that the end-point of the two-segment robot manipulator bumped into an obstacle (Fig. 5.5). In this way the robot is producing a force against the obstacle. The horizontal component of the force acts in the positive direction of the x axis, while the vertical component is directed along the y axis. The force against the obstacle is produced by the motors in the robot joints. The motor of the first joint is producing the torque M_1, while M_2 is the torque in the second joint.

The positive directions of both joint torques are counter-clockwise. As the robot is not moving, the sum of the external torques equals zero. This means that the torque M_1 in the first joint is equal to the torque of the external force or it is equal to the torque that the manipulator exerts on the obstacle

$$M_1 = -F_x\, y + F_y\, x. \tag{5.16}$$

The end-point coordinates x and y, calculated by Eq. (5.4), are inserted into Eq. (5.16)

$$M_1 = -F_x(l_1 \sin\vartheta_1 + l_2 \sin(\vartheta_1 + \vartheta_2)) + F_y(l_1 \cos\vartheta_1 + l_2 \cos(\vartheta_1 + \vartheta_2)). \tag{5.17}$$

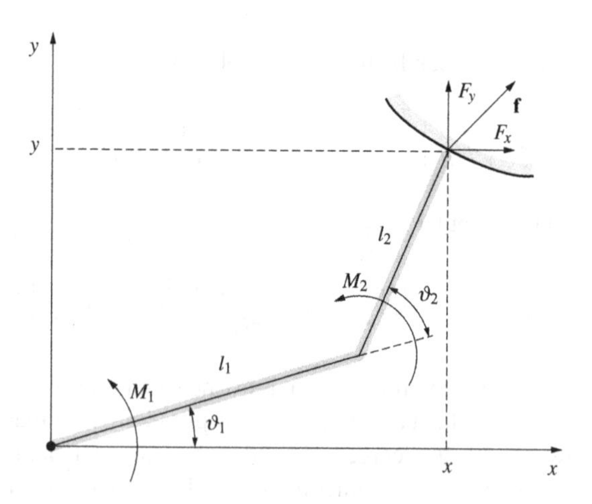

Fig. 5.5 Two-segment robot manipulator in contact with the environment

In a similar way the torque in the second joint is determined

$$M_2 = -F_x l_2 \sin(\vartheta_1 + \vartheta_2) + F_y l_2 \cos(\vartheta_1 + \vartheta_2). \tag{5.18}$$

Equations (5.17) and (5.18) can be written in matrix form

$$\begin{bmatrix} M_1 \\ M_2 \end{bmatrix} = \begin{bmatrix} -l_1\, s1 - l_2\, s12 & l_1\, c1 + l_2\, c12 \\ -l_2\, s12 & l_2\, c12 \end{bmatrix} \begin{bmatrix} F_x \\ F_y \end{bmatrix}. \tag{5.19}$$

The matrix in Eq. (5.19) is a transposed Jacobian matrix. The transposed matrix of order 2×2 has the following form

$$\mathbf{A} = \begin{bmatrix} a & b \\ c & d \end{bmatrix} \quad \mathbf{A}^T = \begin{bmatrix} a & c \\ b & d \end{bmatrix}.$$

In this way we obtained an important relation between the joint torques and the forces at the robot end-effector

$$\tau = \mathbf{J}^T(\mathbf{q})\mathbf{f}, \tag{5.20}$$

where

$$\tau = \begin{bmatrix} M_1 \\ M_2 \end{bmatrix} \qquad \mathbf{f} = \begin{bmatrix} F_x \\ F_y \end{bmatrix}.$$

Equation (5.20) describes the robot statics. It will be used in the control of a robot which is in contact with the environment.

5.3 Workspace

The robot workspace consists of all points that can be reached by the robot end-point. It plays an important role when selecting an industrial robot for an anticipated task. It is our aim to describe an approach to determine the workspace of a chosen robot. We shall again consider the example of the simple planar two-segment robot with rotational joints. Our study of the robot workspace will therefore take place in a plane and we shall in fact deal with a working surface. Regardless of the constraints imposed by the plane we shall become aware of the most important characteristic properties of the robot workspaces. Industrial robots usually have the ability to rotate around the first vertical joint axis. We shall therefore rotate the working surface around the vertical axis of the reference coordinate frame and thus obtain an idea of the realistic three-dimensional robot workspaces.

Let us consider the planar two-segment robot manipulator as shown in Fig. 5.6. The rotational degrees of freedom are denoted as ϑ_1 and ϑ_2. The lengths of the segments l_1 and l_2 will be considered equal. The coordinates of the robot end-point can be expressed as in (5.4) with the following two equations:

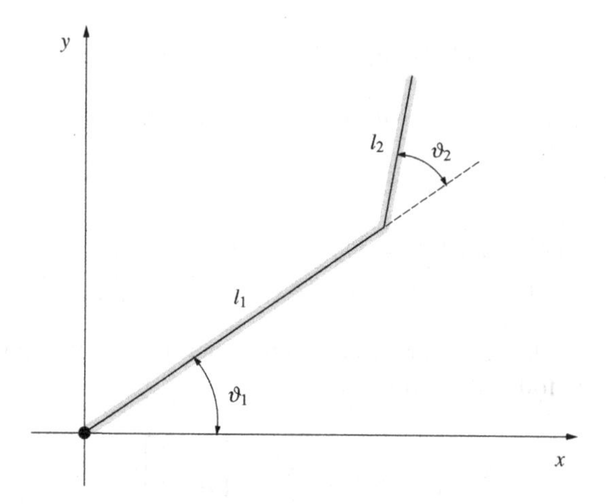

Fig. 5.6 Two-segment robot manipulator

$$\begin{aligned} x &= l_1 \cos \vartheta_1 + l_2 \cos(\vartheta_1 + \vartheta_2) \\ y &= l_1 \sin \vartheta_1 + l_2 \sin(\vartheta_1 + \vartheta_2). \end{aligned} \tag{5.21}$$

If Eqs. (5.21) are first squared and then summed, the equations of a circle are obtained

$$\begin{aligned} (x - l_1 \cos \vartheta_1)^2 + (y - l_1 \sin \vartheta_1)^2 &= l_2^2 \\ x^2 + y^2 &= l_1^2 + l_2^2 + 2l_1 l_2 \cos \vartheta_2. \end{aligned} \tag{5.22}$$

The first equation depends only on the angle ϑ_1, while only ϑ_2 appears in the second equation. The mesh of the circles plotted for different values ϑ_1 and ϑ_2 is shown in Fig. 5.7. The first equation describes the circles which are in Fig. 5.7 denoted as $\vartheta_1 = 0°, 30°, 60°, 90°, 120°, 150°$, and $180°$. Their radii are equal to the length of the second segment l_2, the centers of the circles depend on the angle ϑ_1 and travel along a circle with the center in the origin of the coordinate frame and with the radius l_1. The circles of the second equation have all their centers in the origin of the coordinate frame, while their radii depend on the lengths of both segments and the angle ϑ_2 between them.

The mesh in Fig. 5.7 serves for a simple graphical presentation of the working surface of a two-segment robot. It is not difficult to determine the working surface for the case when ϑ_1 and ϑ_2 vary in the full range from 0° to 360°. For the two-segment manipulator with equal lengths of both segments this is simply a circle with the radius $l_1 + l_2$. Much more irregular shapes of workspaces are obtained when the range of motion of the robot joints is constrained, as is usually the case. Part of the working surface where ϑ_1 changes from 0° to 60° and ϑ_2 from 60° to 120° is displayed as hatched in Fig. 5.7.

When plotting the working surfaces of the two-segment manipulator we assumed that the lengths of both segments are equal. This assumption will be now supported by

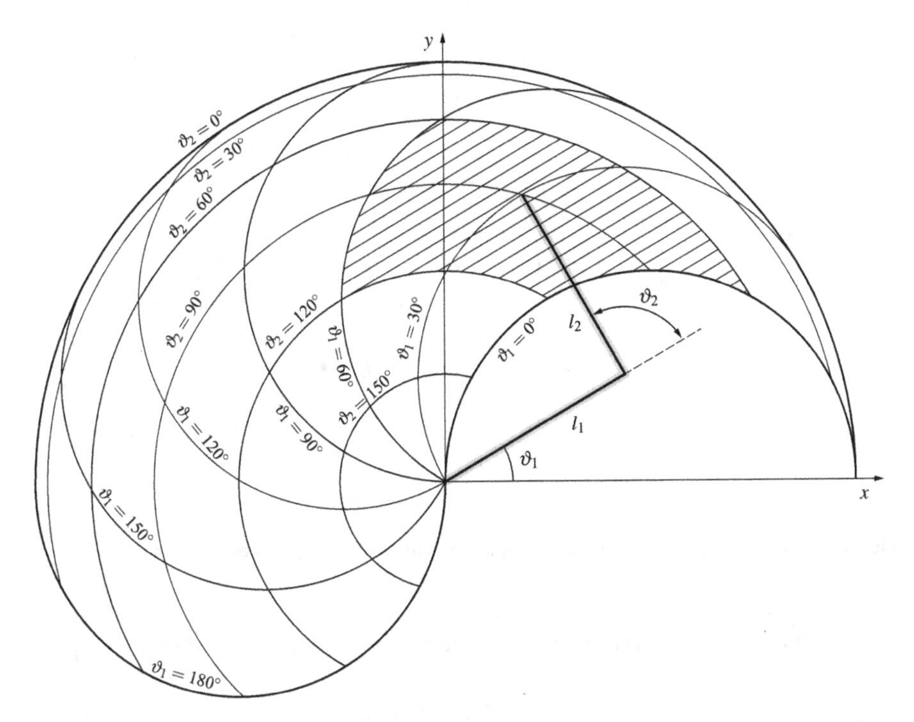

Fig. 5.7 Workspace of a planar two-segment robot manipulator ($l_1 = l_2$, $0° \leq \vartheta_1 \leq 180°$, $0° \leq \vartheta_2 \leq 180°$)

an adequate proof. It is not difficult to realize that the segments of industrial SCARA and anthropomorphic robots are of equal length. Let us consider a two-segment robot, where the second segment is shorter than the first one, while the angles ϑ_1 and ϑ_2 vary from 0° to 360° (Fig. 5.8). The working area of such a manipulator is a ring with inner radius $R_i = l_1 - l_2$ and outer radius $R_o = l_1 + l_2$. It is our aim to find the ratio of the segments lengths l_1 and l_2 resulting in the largest working area at constant sum of lengths of both segments R_o. The working area of the described two-segment robot manipulator is

$$A = \pi R_o^2 - \pi R_i^2. \tag{5.23}$$

By inserting the expression for the inner radius in Eq. (5.23)

$$R_i^2 = (l_1 - l_2)^2 = (2l_1 - R_o)^2 \tag{5.24}$$

we can write

$$A = \pi R_o^2 - \pi (2l_1 - R_o)^2. \tag{5.25}$$

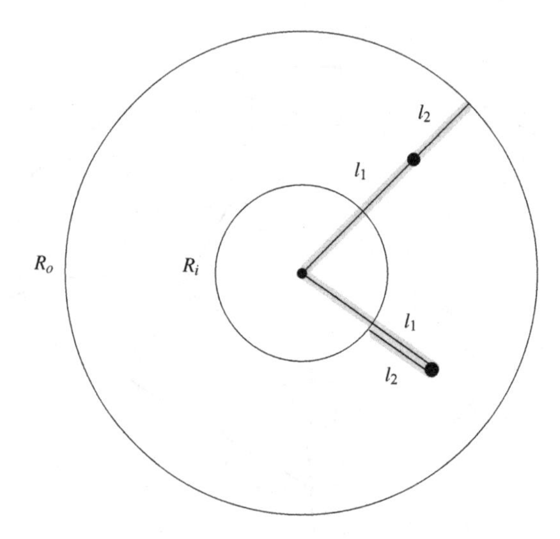

Fig. 5.8 Working area of two-segment manipulator with the second segment shorter

For maximum area, the derivative with respect to segment length l_1 should be equal zero

$$\frac{\partial A}{\partial l_1} = 2\pi(2l_1 - R_o) = 0. \tag{5.26}$$

The solution is

$$l_1 = \frac{R_o}{2}, \tag{5.27}$$

giving

$$l_1 = l_2. \tag{5.28}$$

The largest working area of the two-segment mechanism occurs for equal lengths of both segments.

The area of the working surface depends on the segment lengths l_1 and l_2 and on the minimal and maximal values of the angles ϑ_1 and ϑ_2. When changing the ratios l_1/l_2 we can obtain various shapes of the robot working surface. The area of a such working surface is always equal to the one shown in Fig. 5.9. In this Figure $\Delta\vartheta_1$ refers to the difference between the maximal and minimal joint angle value $\Delta\vartheta_1 = (\vartheta_{1_{max}} - \vartheta_{1_{min}})$. The area of the working surface is the area of a ring segment

$$A = \frac{\Delta\vartheta_1 \pi}{360}(r_1^2 - r_2^2) \tag{5.29}$$

for $\Delta\vartheta_1$ given in angular degrees.

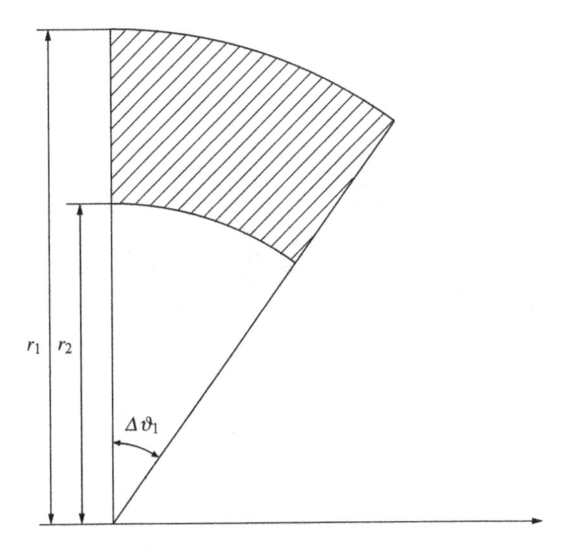

Fig. 5.9 Working surface of a two-segment manipulator

In Eq. (5.29), the radii r_1 and r_2 are obtained by the cosine rule

$$r_1 = \sqrt{l_1^2 + l_2^2 + 2l_1l_2\cos\vartheta_{2_{min}}} \qquad r_2 = \sqrt{l_1^2 + l_2^2 + 2l_1l_2\cos\vartheta_{2_{max}}}. \tag{5.30}$$

The area of the working surface is, in the same way as its shape, dependent on the ratio l_2/l_1 and on the constraints in the joint angles. The angle ϑ_1 determines the position of the working surface with respect to the reference frame and has no influence on its shape. Let us examine the influence of the second angle ϑ_2 on the area of the working surface. We shall assume that $l_1 = l_2 = 1$ and ϑ_1 change from 30° to 60°. For equal ranges of the angle ϑ_2 (30°) and for different values of $\vartheta_{2_{max}}$ and $\vartheta_{2_{min}}$ we obtain different values of the working areas

$$\begin{aligned} 0^\circ \le \vartheta_2 \le 30^\circ \quad & A = 0.07 \\ 30^\circ \le \vartheta_2 \le 60^\circ \quad & A = 0.19 \\ 60^\circ \le \vartheta_2 \le 90^\circ \quad & A = 0.26 \\ 90^\circ \le \vartheta_2 \le 120^\circ \quad & A = 0.26 \\ 120^\circ \le \vartheta_2 \le 150^\circ \quad & A = 0.19 \\ 150^\circ \le \vartheta_2 \le 180^\circ \quad & A = 0.07. \end{aligned}$$

Until now, under the term workspace we were considering the so called reachable robot workspace. This includes all the points in the robot surroundings that can be reached by the robot end-point. Often this so-called dexterous workspace is of greater importance. The dexterous workspace comprises all the points that can be reached with any arbitrary orientation of the robot end-effector. This workspace is always

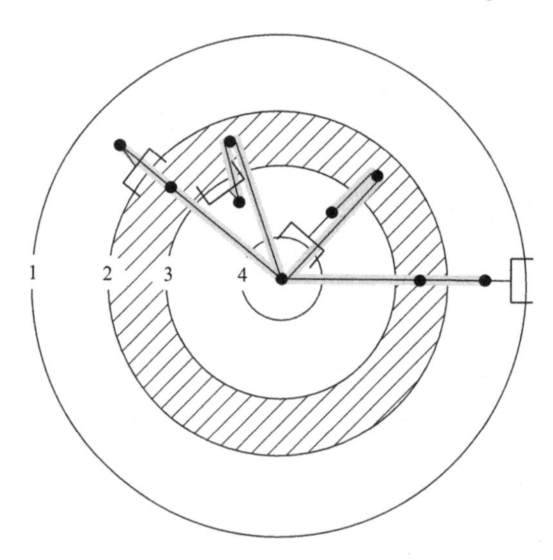

Fig. 5.10 Reachable and dexterous workspace of a two-segment manipulator with end-effector

smaller than the reachable workspace. The dexterous workspace is larger when the last segment (end-effector) is shorter. The reachable and the dexterous workspaces of a two-segment robot with the end-effector are shown in Fig. 5.10. The second and the third circle are obtained when the robot end-effector is oriented towards the area constrained by the two circles. These two circles represent the limits of the dexterous workspace. The first and the fourth circle constrain the reachable workspace. The points between the first and the second and the third and the fourth circle cannot be reached with an arbitrary orientation of the end-effector.

For robots having more than three joints, the described graphical approach is not appropriate. In that case we make use of numerical methods and computer algorithms.

5.4 Dynamics

For illustration purposes, we shall study the planar, two-segment robot manipulator as shown in Fig. 5.11. The segments of length l_1 and l_2 may move in the vertical x–y plane, their positions being described by angles with respect to the horizontal (x) axis; ϑ_1 and $\vartheta = \vartheta_1 + \vartheta_2$. Actuators at the joints provide torques M_1 and M_2, whose positive direction is defined by increasing angles, i.e., along the positive direction of the z axis of our reference coordinate frame.

We now approximate the segments by point masses m_1 and m_2 at the midpoints of rigid, but otherwise massless rods (see Fig. 5.12). Let $\mathbf{r}_1$ denote the position of point mass m_1 with respect to the first joint, which is at the origin of our coordinate frame. Let $\mathbf{r}_2$ denote the position of point mass m_2 with respect to the second joint, which is at the junction of the two segments.

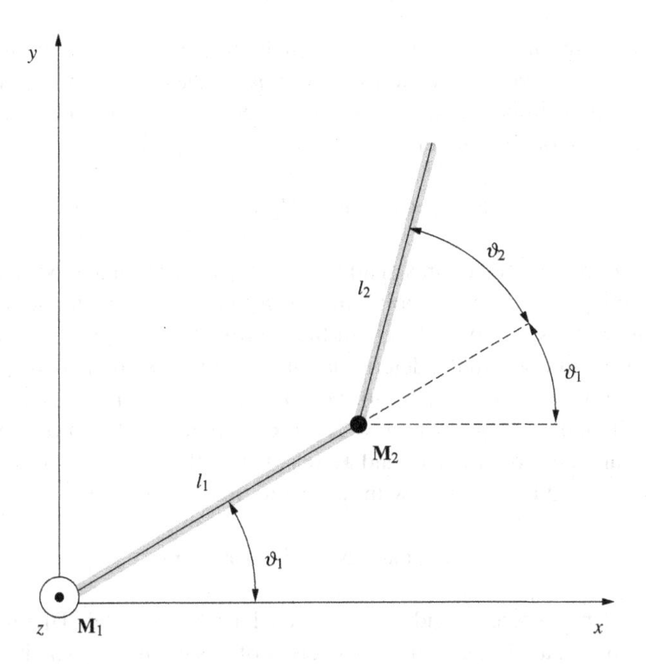

Fig. 5.11 Parameters of the planar, two-segment robot manipulator, which moves in the vertical x–y plane

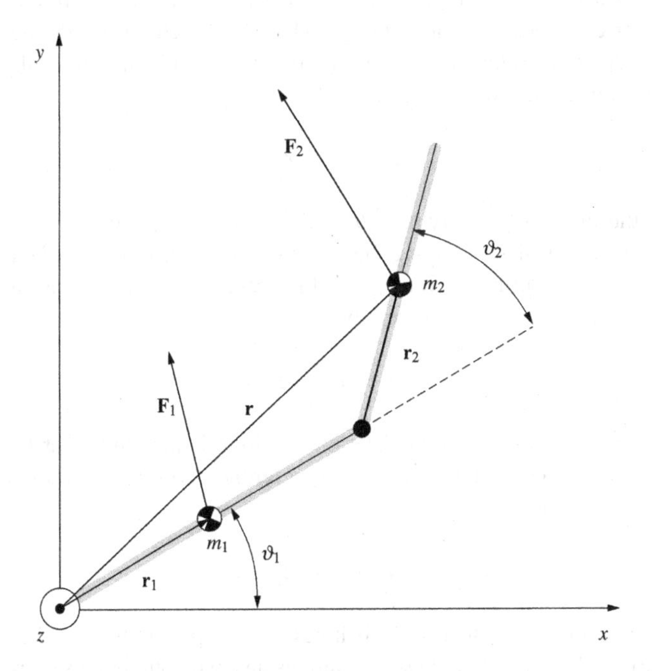

Fig. 5.12 Planar robot manipulator approximated by point masses m_1 and m_2

The point masses m_1 and m_2 are acted upon by the forces that are transmitted by the massless rods, as well as by the force of gravity. Newton's law claims that the vector sum of all the forces acting on a particle is equal to the product of the particle mass and it's acceleration. Therefore,

$$\mathbf{F}_1 = m_1\mathbf{a}_1 \quad \text{and} \quad \mathbf{F}_2 = m_2\mathbf{a}_2, \tag{5.31}$$

where $\mathbf{F}_1$ and $\mathbf{F}_2$ represent the sums of all forces (i.e., the force of the rod and the force of gravity), acting on each of the point masses m_1 and m_2, while $\mathbf{a}_1$ and $\mathbf{a}_2$ are their accelerations with respect to the origin of the coordinate frame. So, a calculation of the accelerations amounts to the determination of the forces on the two "particles".

The position of m_1 with respect to the reference frame origin, is given by $\mathbf{r_1}$, while the position of m_2 is given by $\mathbf{r} = 2\mathbf{r_1} + \mathbf{r_2}$ (see Fig. 5.12). The corresponding accelerations are therefore $\mathbf{a}_1 = \ddot{\mathbf{r}}_1$ and $\mathbf{a}_2 = \ddot{\mathbf{r}}$, where the two dots above the vector symbol denote second derivatives with respect to time. Therefore,

$$\mathbf{a}_1 = \ddot{\mathbf{r}}_1 \quad \text{and} \quad \mathbf{a}_2 = \ddot{\mathbf{r}} = 2\ddot{\mathbf{r}}_1 + \ddot{\mathbf{r}}_2. \tag{5.32}$$

Now, $\mathbf{r}_1$ and $\mathbf{r}_2$ represent rigid rods, so their lengths are fixed. Therefore, these vectors can only rotate. Let us remind ourselves of basic physics, which says that a rotating vector describes a particle in circular motion. Such motion may have two components of acceleration (Fig. 5.13, see also Appendix ??). The first component is the radial or centripetal acceleration $\mathbf{a}_r$, which is directed towards the center of rotation. It is due to the change only of the direction of velocity and is thus present also in uniform circular motion. It is given by the expression

$$\mathbf{a}_r = -\omega^2\mathbf{r}, \tag{5.33}$$

where ω is the angular velocity $\omega = \dot{\theta}$. The second component is the tangential acceleration, which is directed along the tangent to the circle (Fig. 5.13). It is due to the change of the magnitude of velocity and is present only in circular motion with angular acceleration $\alpha = \ddot{\theta}$. It is given by

$$\mathbf{a}_t = \boldsymbol{\alpha} \times \mathbf{r}, \tag{5.34}$$

where $\boldsymbol{\alpha}$ is the vector of angular acceleration, which is perpendicular to the plane of motion, i.e., it is along the z axis of our reference coordinate frame. The total acceleration is obviously

$$\mathbf{a} = \mathbf{a}_r + \mathbf{a}_t = -\omega^2\mathbf{r} + \boldsymbol{\alpha} \times \mathbf{r}. \tag{5.35}$$

Let us now calculate the second derivatives with respect to time of the vectors $\mathbf{r}_1$ and $\mathbf{r}_2$. As noted above, each of these derivatives has two components corresponding to the radial and to the tangential acceleration. So

$$\ddot{\mathbf{r}}_1 = -\omega_1^2\mathbf{r}_1 + \boldsymbol{\alpha}_1 \times \mathbf{r}_1 \quad \text{and} \quad \ddot{\mathbf{r}}_2 = -\omega_2^2\mathbf{r}_2 + \boldsymbol{\alpha}_2 \times \mathbf{r}_2. \tag{5.36}$$

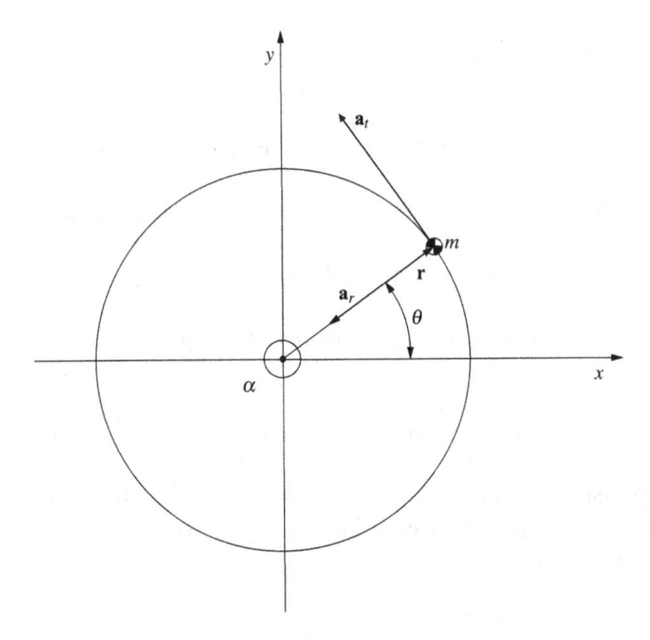

Fig. 5.13 A rotating vector **r** of fixed length describes a particle in circular motion

The magnitude of the angular velocity ω_1 and the vector of angular acceleration $\boldsymbol{\alpha}_1$ of the first segment are

$$\omega_1 = \dot{\vartheta}_1 \quad \text{and} \quad \boldsymbol{\alpha}_1 = \ddot{\vartheta}_1 \mathbf{k}, \tag{5.37}$$

where **k** is the unit vector along the z axis. The angular velocity ω_2 and the angular acceleration $\boldsymbol{\alpha}_2$ of the second segment are

$$\omega_2 = \dot{\vartheta} = \dot{\vartheta}_1 + \dot{\vartheta}_2 \quad \text{and} \quad \boldsymbol{\alpha}_2 = \ddot{\vartheta}\mathbf{k} = (\ddot{\vartheta}_1 + \ddot{\vartheta}_2)\mathbf{k}. \tag{5.38}$$

Here we used $\vartheta = \vartheta_1 + \vartheta_2$ (see Figs. 5.11 and 5.12). The second derivatives of vectors $\mathbf{r}_1$ and $\mathbf{r}_2$ may be written as

$$\ddot{\mathbf{r}}_1 = -\omega_1^2 \mathbf{r}_1 + \boldsymbol{\alpha}_1 \times \mathbf{r}_1 = -\dot{\vartheta}_1^2 \mathbf{r}_1 + \ddot{\vartheta}_1 \mathbf{k} \times \mathbf{r}_1 \tag{5.39}$$

and

$$\begin{aligned} \ddot{\mathbf{r}}_2 &= -\omega_2^2 \mathbf{r}_2 + \boldsymbol{\alpha}_2 \times \mathbf{r}_2 = -\dot{\vartheta}^2 \mathbf{r}_2 + \ddot{\vartheta}\mathbf{k} \times \mathbf{r}_2 = \\ &= -(\dot{\vartheta}_1 + \dot{\vartheta}_2)^2 \mathbf{r}_2 + (\ddot{\vartheta}_1 + \ddot{\vartheta}_2)\mathbf{k} \times \mathbf{r}_2. \end{aligned} \tag{5.40}$$

We may now use these expressions to calculate the accelerations of the two point masses m_1 and m_2 corresponding to our two-segment robot. The acceleration $\mathbf{a}_1$ of m_1 is

$$\mathbf{a}_1 = \ddot{\mathbf{r}}_1 = -\dot{\vartheta}_1^2 \mathbf{r}_1 + \ddot{\vartheta}_1 (\mathbf{k} \times \mathbf{r}_1). \tag{5.41}$$

The acceleration $\mathbf{a}_2$ of m_2 is

$$\begin{aligned}\mathbf{a}_2 &= \ddot{\mathbf{r}} = 2\ddot{\mathbf{r}}_1 + \ddot{\mathbf{r}}_2 = \\ &= -2\dot{\vartheta}_1^2\mathbf{r}_1 + 2\ddot{\vartheta}_1(\mathbf{k}\times\mathbf{r}_1) - (\dot{\vartheta}_1+\dot{\vartheta}_2)^2\mathbf{r}_2 + (\ddot{\vartheta}_1+\ddot{\vartheta}_2)(\mathbf{k}\times\mathbf{r}_2).\end{aligned} \tag{5.42}$$

From these accelerations we get the total forces acting on particles m_1 and m_2

$$\mathbf{F}_1 = m_1\mathbf{a}_1 \quad \text{and} \quad \mathbf{F}_2 = m_2\mathbf{a}_2. \tag{5.43}$$

We can now calculate the torques of these forces with respect to the coordinate frame origin

$$\boldsymbol{\tau}_1 = \mathbf{r}_1 \times \mathbf{F}_1 = \mathbf{r}_1 \times m_1\mathbf{a}_1 \quad \text{and} \quad \boldsymbol{\tau}_2 = \mathbf{r} \times \mathbf{F}_2 = (2\mathbf{r}_1 + \mathbf{r}_2) \times m_2\mathbf{a}_2. \tag{5.44}$$

Inserting expressions for $\mathbf{a}_1$ and $\mathbf{a}_2$ as derived above, reminding ourselves of the double vector product $[\mathbf{a}\times(\mathbf{b}\times\mathbf{c}) = \mathbf{b}(\mathbf{a}\cdot\mathbf{c}) - \mathbf{c}(\mathbf{a}\cdot\mathbf{b})]$, and by patiently doing the lengthy algebra, we obtain

$$\begin{aligned}\boldsymbol{\tau}_1 =& m_1 r_1^2\ddot{\vartheta}_1\mathbf{k} \\ &\text{and} \\ \boldsymbol{\tau}_2 =& [\ddot{\vartheta}_1(4m_2r_1^2 + 4m_2r_1r_2\cos\vartheta_2 + m_2r_2^2) + \\ &+ \ddot{\vartheta}_2(m_2r_2^2 + 2m_2r_1r_2\cos\vartheta_2) - \\ &- \dot{\vartheta}_1\dot{\vartheta}_2 4m_2r_1r_2\sin\vartheta_2 - \dot{\vartheta}_2^2 2m_2r_1r_2\sin\vartheta_2]\mathbf{k}.\end{aligned} \tag{5.45}$$

The sum of both torques on the two "particles" of our system is obviously $\boldsymbol{\tau} = \boldsymbol{\tau}_1 + \boldsymbol{\tau}_2$.

On the other hand, we may consider our two-segment system consisting of two point masses and two massless rods from a different viewpoint. As a consequence of Newton's third law (To every action there is an equal but opposite reaction), we have a theorem stating that internal torques in a system cancel out, so that only torques of external forces are relevant. The torques of external forces on our robot system are the torques of gravity and the torque exerted by the base on which the robot stands. The torque of the base is equal to the torque $\mathbf{M}_1$ of the actuator in the first joint. The sum of these torques of external forces (base + gravity) must be equal to $\boldsymbol{\tau}_1 + \boldsymbol{\tau}_2$ (derived above), as both results represent two ways of viewing the total torque on the same system. So

$$\mathbf{M}_1 + \mathbf{r}_1 \times m_1\mathbf{g} + \mathbf{r} \times m_2\mathbf{g} = \boldsymbol{\tau}_1 + \boldsymbol{\tau}_2. \tag{5.46}$$

With $\mathbf{r} = 2\mathbf{r}_1 + \mathbf{r}_2$ we have the torque of the actuator in the first joint

$$\mathbf{M}_1 = \boldsymbol{\tau}_1 + \boldsymbol{\tau}_2 - \mathbf{r}_1 \times m_1\mathbf{g} - (2\mathbf{r}_1 + \mathbf{r}_2) \times m_2\mathbf{g}. \tag{5.47}$$

Remembering that $\mathbf{g}$ points vertically downward (i.e., along $-y$) and inserting the above results for $\boldsymbol{\tau}_1$ and $\boldsymbol{\tau}_2$, we finally get

$$\begin{aligned} M_1 = {} & \ddot{\vartheta}_1(m_1 r_1^2 + m_2 r_2^2 + 4 m_2 r_1^2 + 4 m_2 r_1 r_2 \cos\vartheta_2) + \\ & + \ddot{\vartheta}_2(m_2 r_2^2 + 2 m_2 r_1 r_2 \cos\vartheta_2) - \\ & - \dot{\vartheta}_1 \dot{\vartheta}_2 4 m_2 r_1 r_2 \sin\vartheta_2 - \dot{\vartheta}_2^2 2 m_2 r_1 r_2 \sin\vartheta_2 + \\ & + m_1 g r_1 \cos\vartheta_1 + 2 m_2 g r_1 \cos\vartheta_1 + m_2 g r_2 \cos(\vartheta_1 + \vartheta_2). \end{aligned} \tag{5.48}$$

In order to obtain the torque $\mathbf{M}_2$ of the actuator in the second joint, we will first consider the total force $\mathbf{F}_2$ acting on the point mass m_2. The force $\mathbf{F}_2$ is a sum of two contributions. One is the force of gravity $m_2\mathbf{g}$, the other is the force $\mathbf{F}_2'$ exerted on m_2 by the massless and rigid rod of the second segment. So

$$\mathbf{F}_2 = \mathbf{F}_2' + m_2\mathbf{g}. \tag{5.49}$$

To this equation we apply a vector product of $\mathbf{r}_2$ from the left and obtain

$$\mathbf{r}_2 \times \mathbf{F}_2 = \mathbf{r}_2 \times \mathbf{F}_2' + \mathbf{r}_2 \times m_2\mathbf{g}. \tag{5.50}$$

The first term on the right-hand side is the vector product of $\mathbf{r}_2$ with the force $\mathbf{F}_2'$ exerted on m_2 by the massless and rigid rod. This term is equal to the torque $\mathbf{M}_2$ of the actuator in the second joint. (Note that the rod may also exert a force on m_2 directed along the rod, but the vector product of that component with $\mathbf{r}_2$ vanishes). We therefore obtain

$$\mathbf{M}_2 = \mathbf{r}_2 \times \mathbf{F}_2 - \mathbf{r}_2 \times m_2\mathbf{g}. \tag{5.51}$$

Substituting $m_2\mathbf{a}_2$ for $\mathbf{F}_2$ and the expression derived previously for $\mathbf{a}_2$, leads to

$$\begin{aligned} M_2 = {} & \ddot{\vartheta}_1(m_2 r_2^2 + 2 m_2 r_1 r_2 \cos\vartheta_2) + \ddot{\vartheta}_2 m_2 r_2^2 + \\ & + \dot{\vartheta}_1^2 2 m_2 r_1 r_2 \sin\vartheta_2 + m_2 r_2 g \cos(\vartheta_1 + \vartheta_2). \end{aligned} \tag{5.52}$$

The expressions for M_1 (5.48) and M_2 (5.52) seem relatively complicated, so let us investigate some simple and familiar cases. First, assume $\vartheta_1 = -90°$ and no torque in the second joint $M_2 = 0$ (Fig. 5.14 left). The equation for M_2 reduces to

$$\ddot{\vartheta}_2 m_2 r_2^2 = -m_2 g r_2 \sin\vartheta_2. \tag{5.53}$$

This is the equation of a simple pendulum with mass m_2, moment of inertia $m_2 r_2^2 = J_2$, which is rotating around the second joint with angular acceleration $\ddot{\vartheta}_2$ (Fig. 5.14 left). The left-hand side is thus $J_2\alpha_2$ and on the right-hand side we have the torque due to gravity. So, this is an example of the simple equation $M = J\alpha$, to which our

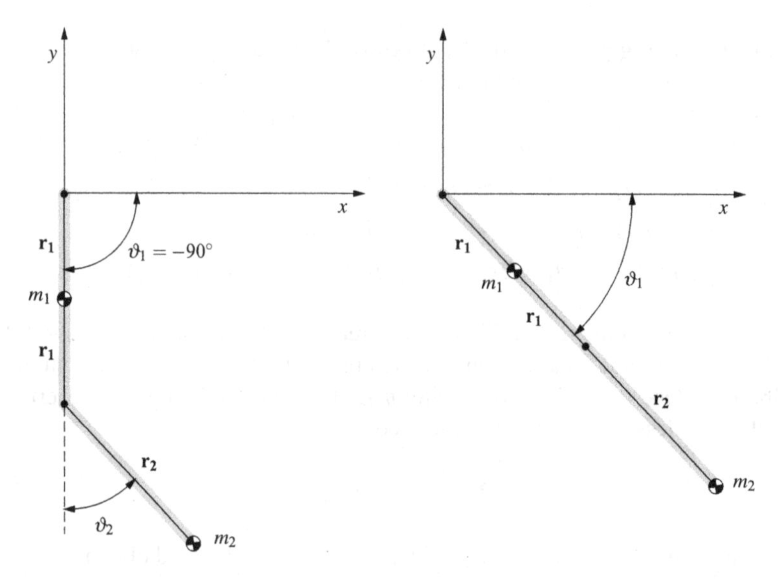

Fig. 5.14 Two simple examples of the two-segment robot manipulator: $\vartheta_1 = -90°$ (left) and $\vartheta_2 = 0°$ (right)

complicated expression has been reduced. For small oscillations ($\vartheta_2 \ll 1$) we have $\sin \vartheta_2 \approx \vartheta_2$ and the equation becomes

$$\ddot{\vartheta}_2 + (\frac{g}{r_2})\vartheta_2 = 0. \tag{5.54}$$

This is the equation of the simple pendulum with angular frequency $\omega_0 = \sqrt{\frac{g}{r_2}}$ and oscillation period $T = 2\pi\sqrt{\frac{r_2}{g}}$.

Next assume $\vartheta_2 = 0$ so we have one rigid rod rotating around one end, which is fixed at the coordinate frame origin (Fig. 5.14 right). If we also "switch off" gravity ($g = 0$), we obtain for the torque in the first joint

$$\begin{aligned} M_1 &= \ddot{\vartheta}_1(m_1 r_1^2 + m_2 r_2^2 + 4m_2 r_1^2 + 4m_2 r_1 r_2) = \\ &= \ddot{\vartheta}_1[m_1 r_1^2 + m_2(2r_1 + r_2)^2] = J_{12}\alpha_1, \end{aligned} \tag{5.55}$$

where $\alpha_1 = \ddot{\vartheta}_1$ is the angular acceleration and J_{12} is the combined moment of inertia of the two masses. Alternatively, one might take the torque in the first joint equal to zero, include gravity and one gets a relatively simple pendulum with two point masses on one massless rigid rod.

Let us mention that the above full equations for M_1 and M_2 (Eqs. (5.48) and (5.52)), with minor adaptations of notation, are valid for a double pendulum with

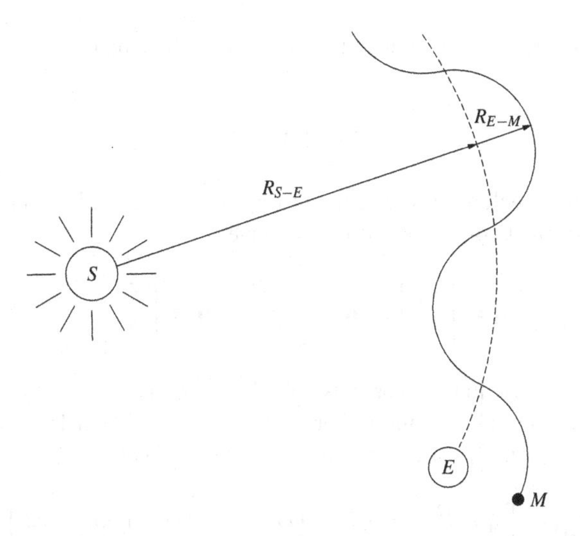

Fig. 5.15 Schematic trajectory (not in scale) of the Earth (dashed curve) and the Moon (full curve) in the reference frame of the Sun

friction. The torques of the actuators are in this case replaced by the torques of friction in the joints.

An amusing exercise would be to compare the trajectories of the endpoints of the two segments of our simple robot with the trajectories of the Earth and the Moon, as seen from the reference frame of the Sun. Let us approximate the Earth and the Moon as point particles ($m_E \gg m_M$) in coplanar circular orbits. As the gravitational force acts only along the line joining the two particles, it cannot transmit torques, so the angular accelerations are zero and the angular velocities are constant. The orbital velocity of the Earth around the Sun ($R_{S-E} \approx 150 \cdot 10^6$ km, $T = 365$ days, $v_E \approx 2.6 \cdot 10^6$ km/day) is much greater than the orbital velocity of the Moon around the Earth ($R_{E-M} \approx 0.38 \cdot 10^6$ km, $T = 28$ days, $v_M \approx 0.08 \cdot 10^6$ km/day), so the trajectory of the Moon as seen in the Sun's reference frame would be approximately a sine curve superimposed on the Earth's circular orbit around the Sun (Fig. 5.15). With our two-segment robot one could have higher angular velocities of the second segment resulting in different shapes of the trajectory of its endpoint (Ptolemy's epicycles for example).

Returning to our relatively complicated equations for the torques M_1 and M_2 (Eqs. (5.48) and (5.52)), due to actuators in the joints, we see that these equations may be condensed into matrix form representing the robot dynamic model as

$$\tau = B(\mathbf{q})\ddot{\mathbf{q}} + C(\mathbf{q}, \dot{\mathbf{q}})\dot{\mathbf{q}} + \mathbf{g}(\mathbf{q}). \tag{5.56}$$

In this equation the vector $\boldsymbol{\tau}$ unites the torques of both actuators

$$\boldsymbol{\tau} = \begin{bmatrix} M_1 \\ M_2 \end{bmatrix}. \tag{5.57}$$

Vectors $\mathbf{q}$, $\dot{\mathbf{q}}$ and $\ddot{\mathbf{q}}$ belong to the segment trajectories, velocities and accelerations respectively. For the two-segment robot we have

$$\mathbf{q} = \begin{bmatrix} \vartheta_1 \\ \vartheta_2 \end{bmatrix}, \quad \dot{\mathbf{q}} = \begin{bmatrix} \dot{\vartheta}_1 \\ \dot{\vartheta}_2 \end{bmatrix}, \quad \ddot{\mathbf{q}} = \begin{bmatrix} \ddot{\vartheta}_1 \\ \ddot{\vartheta}_2 \end{bmatrix}.$$

The first term of the equation for $\boldsymbol{\tau}$ is called the inertial term. In our case of the planar, two-segment robot manipulator with $r_1 = r_2 = \frac{l}{2}$ and by simplifying the notation with $s1 = \sin\vartheta_1$, $c12 = \cos(\vartheta_1 + \vartheta_2)$ etc., we get

$$\mathbf{B}(\mathbf{q}) = \begin{bmatrix} \frac{1}{4}m_1l^2 + \frac{5}{4}m_2l^2 + m_2l^2c2 & \frac{1}{4}m_2l^2 + \frac{1}{2}m_2l^2c2 \\ \frac{1}{4}m_2l^2 + \frac{1}{2}m_2l^2c2 & \frac{1}{4}m_2l^2 \end{bmatrix}. \tag{5.58}$$

The second term of this matrix equation is called the Coriolis term and includes velocity and centrifugal effects. For the two-segment robot we have the following matrix

$$\mathbf{C}(\mathbf{q}, \dot{\mathbf{q}}) = \begin{bmatrix} -m_2l^2s2\dot{\vartheta}_2 & -\frac{1}{2}m_2l^2s2\dot{\vartheta}_2 \\ \frac{1}{2}m_2l^2s2\dot{\vartheta}_1 & 0 \end{bmatrix}. \tag{5.59}$$

The gravitational column $\mathbf{g}(\mathbf{q})$ has in our case the following form

$$\mathbf{g}(\mathbf{q}) = \begin{bmatrix} \frac{1}{2}m_1glc1 + m_2glc1 + \frac{1}{2}m_2glc12 \\ \frac{1}{2}m_2glc12 \end{bmatrix}. \tag{5.60}$$

Chapter 6
Parallel Robots

This chapter deals with the increasingly popular and high-performing robots that are known as parallel robots. Standard mechanisms of industrial robots possess serial kinematic chains in which links and joints alternate as shown in Fig. 6.1 (left). These are referred to as serial robots. Lately, we have seen a significant advancement of parallel robots. They include closed kinematic chains, an example is shown in Fig. 6.1 (right).

In industry, parallel robots have started to gain ground in the last two decades. However, the initial developments date back to 1962 when Gough and Whitehall developed a parallel robot for testing automobile tires. At about the same time, a similar parallel robot was introduced by Stewart to design a flight simulator. The parallel robot, in which a mobile platform is controlled by six actuated legs, is therefore called the Stewart-Gough platform. The breakthrough of parallel robots was also largely due to the robot developed by Clavel in the eighties. His mechanism was patented in the USA in 1990 under the name of the Delta robot. The parallel mechanisms in robotics had become a subject of systematic scientific research in the early eighties. These activities intensified significantly in the nineties and culminated with some key achievements in robot kinematics in general.

6.1 Characteristics of Parallel Robots

In serial robots, the number of degrees of freedom is identical to the total number of degrees of freedom in joints. Thus, all joints must be actuated, and usually only simple one degree of freedom translational and rotational joints are used. In parallel robots, the number of degrees of freedom is lower than the total number of degrees of freedom in joints so that many joints are passive. Passive joints can be more complex; typical representatives are the universal joint and the spherical joint. The universal joint consists of two perpendicular rotations while three perpendicular

M. Mihelj et al., *Robotics*, https://doi.org/10.1007/978-3-319-72911-4_6

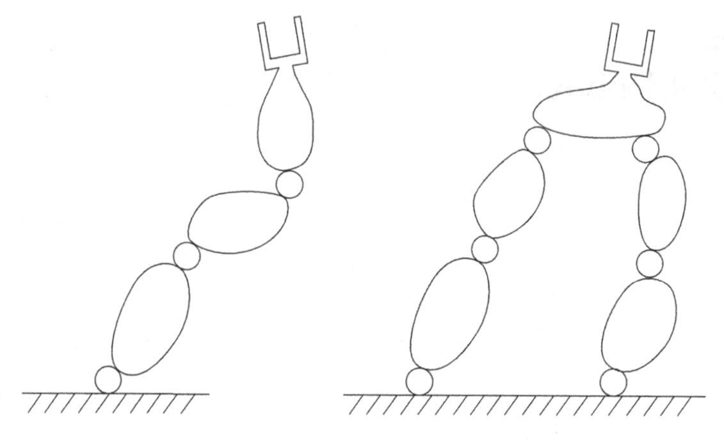

Fig. 6.1 Serial kinematic chain (left) and closed kinematic chain (right)

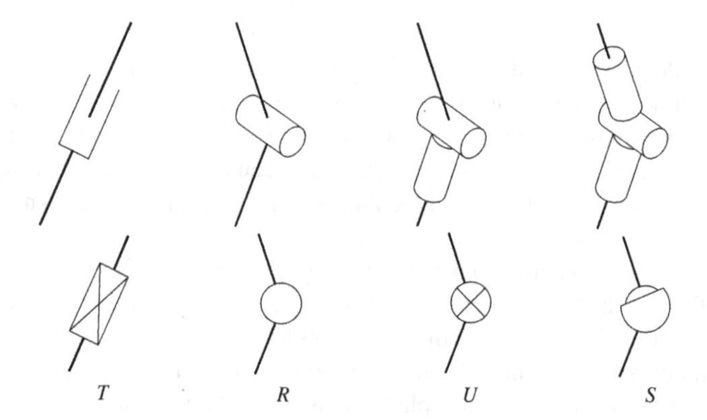

Fig. 6.2 Types of joints often used in parallel mechanisms

rotations compose the spherical joint as shown in Fig. 6.2. Here, letters T, R, U, and S are used to mark the translational joint, the rotational joint, the universal joint, and the spherical joint, respectively.

In parallel robots, the last (top) link of the mechanisms is the so called platform (Fig. 6.3). The platform is the active link to which the end-effector is attached. It is connected to the fixed base by a given number of (usually) serial mechanisms called legs. The whole structure contains at least one closed kinematic chain (minimum two legs). The displacements in the legs produce a displacement of the platform as shown in Fig. 6.3. The motions of the platform and the legs are connected by often very complex trigonometric expressions (direct and inverse kinematics) depending on the geometry of the mechanism, on the type of joints, the number of legs and on their kinematic arrangements.

Unfortunately, unique and uniform denominations for parallel robots do not exist. In this work, a parallel robot is denominated by the type of kinematic chains repre-

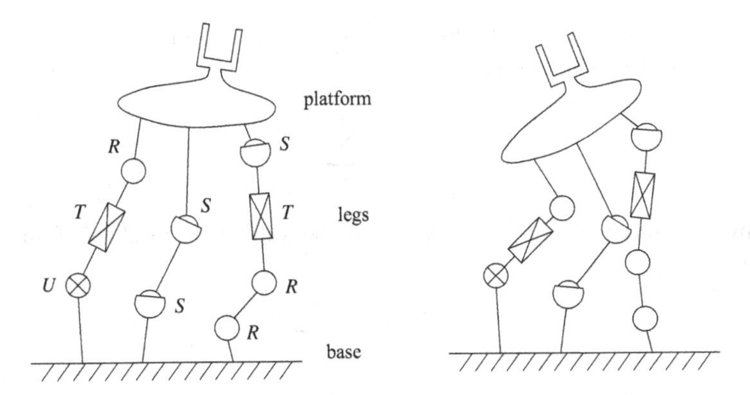

Fig. 6.3 Basic structure of a parallel robot

senting the legs. Thus, the robot in Fig. 6.3 is denominated as UTR-SS-RRTS. When legs of the same type are repeated, for example, in the TRR-TRR-TRR robot, the denomination can be simplified as 3TRR.

Number of degrees of freedom

Each joint contributes to the mobility of the robot by introducing a given number of degrees of freedom or, alternatively, by introducing a corresponding number of constraints, which are defined as follows. Let λ denote the maximum number of degrees of freedom of a freely moving body (in space $\lambda = 6$ and in plane $\lambda = 3$), and let f_i be the number of degrees of freedom of the $i-th$ joint. The corresponding number of constraints is

$$c_i = \lambda - f_i. \tag{6.1}$$

In robotic practice where serial robots dominate, we usually consider joints as elements that add degrees of freedom to the motion of the robot end-effector. In parallel robots, on the contrary, it is more advantageous to consider the movement of the platform (to which the end-effector is attached), taking into account the number of constraints introduced by the joints. Thus, a universal joint U in a space where $\lambda = 6$ introduces $f_i = 2$ degrees of freedom and $c_i = \lambda - f_i = 6 - 2 = 4$ constraints. Or, for example, in a plane where $\lambda = 3$, a rotational joint R introduces $f_i = 1$ degrees of freedom and $c_i = \lambda - f_i = 3 - 1 = 2$ constraints, while the same joint in space introduces $c_i = \lambda - f_i = 6 - 1 = 5$ constraints. Note that rotational and translational joints can operate both in a plane and in space, whereas spherical and universal joints produce only spatial movements and cannot be used in planar robots.

The number of degrees of freedom of a parallel robot is less than the total number of degrees of freedom contributed by the robot joints, unlike in a serial robot where these two numbers are identical. Let N be the number of moving links of the robot and n the number of joints. The joints are referred to as $i = 1, 2, \ldots, n$. Each joint possesses f_i degrees of freedom and c_i constraints. The N free moving links possess $N\lambda$ degrees of freedom. When they are combined into a mechanism, their motion is limited by the constraints introduced by joints, so that the number of degrees of freedom of a robot mechanism is

$$F = N\lambda - \sum_{i=1}^{n} c_i. \tag{6.2}$$

Here, by substituting c_i with $\lambda - f_i$ we obtain the well known Grübler's formula as follows

$$F = \lambda(N - n) + \sum_{i=1}^{n} f_i. \tag{6.3}$$

We must not forget that the number of motors which control the motion of a robot is equal to F.

Note that in serial robots the number of moving links and the number of joints are identical ($N = n$), so that the first part of Grübler's formula is always zero ($\lambda(N - n) = 0$). This explains why the number of degrees of freedom in serial robots is simply

$$F = \sum_{i=1}^{n} f_i. \tag{6.4}$$

A very practical form of Grübler's formula to calculate the degrees of freedom of a parallel robot can be obtained as follows. Suppose that a parallel mechanism includes $k = 1, 2, \ldots, K$ legs, and that each of the legs possesses ν_k degrees of freedom and consequently $\xi_k = \lambda - \nu_k$ constraints. When the platform is not connected to the legs and can freely move in space, it contains λ degrees of freedom. The number of degrees of freedom of a connected platform can thus be computed by subtracting the sum of constraints introduced by the legs

$$F = \lambda - \sum_{k=1}^{K} \xi_k. \tag{6.5}$$

Equations (6.3) and (6.5) are mathematically identical and can be transformed from one to another by simple algebraic operations.

Now we can calculate the degrees of freedom for the robot shown in Fig. 6.3. This robot possesses $N = 7$ moving links and $n = 9$ joints. The total number of degrees of freedom in joints is 16 (3 rotational joints, 2 translational joints, 1 universal and 3 spherical joints). Using the standard Grübler's formula given in Eq. (6.3), we get

$$F = 6(7 - 9) + 16 = 4.$$

If we now use the modified form of Grübler's formula we need to calculate the constraints introduced by each leg. This is rather simple because we only need to subtract the number of degrees of freedom of each leg from λ. For the given robot (legs are counted from left to right) we have $\xi_1 = 2$, $\xi_2 = 0$, and $\xi_3 = 0$. By introducing these values in Eq. (6.5), as expected, we obtain

$$F = 6 - 2 = 4.$$

Advantages and disadvantages of parallel robots

The introduction of parallel robots in industry is motivated by the number of significant advantages that parallel robots have in comparison to serial robots. The most evident are the following:

Load capacity, rigidity, and accuracy. The load carrying capacity of parallel robots is considerably larger than that of serial robots. Parallel robots are also more rigid, and their accuracy in positioning and orienting an end-effector is several times better than with serial robots.

Excellent dynamic properties. The platform can achieve high velocities and accelerations. Furthermore, the resonant frequency of a parallel robot is orders of magnitude higher.

Simple construction. Several passive joints in parallel robots enable less expensive and simple mechanical construction. When building parallel robots standard bearings, spindles, and other machine elements can be used.

The use of parallel robots is, nevertheless, limited. Because of the tangled legs, parallel robots can have difficulties in avoiding obstacles in their workspace. Other significant disadvantages are:

Small workspace. Parallel robots have considerably smaller workspaces than serial robots of comparable size. Their workspace may be further reduced since during motion of the platform the legs may interfere with each other.

Complex kinematics. The computation of kinematics of parallel robots is complex and lengthy. In contrast to serial robots, where the difficulty arises when solving the inverse kinematics problem, in parallel mechanisms the difficulty arises in solving the direct kinematics problem.

Fatal kinematic singularities. Serial robots in kinematically singular poses lose mobility. Parallel robots in singular poses gain degrees of freedom, which cannot be controlled. This is a fatal situation because it cannot be resolved.

6.2 Kinematic Arrangements of Parallel Robots

We can create an immense number of kinematic arrangements of parallel robots. In industrial practice, however, only few of these are used. The most popular and general in the kinematic sense is the Stewart-Gough platform as shown in Fig. 6.4.

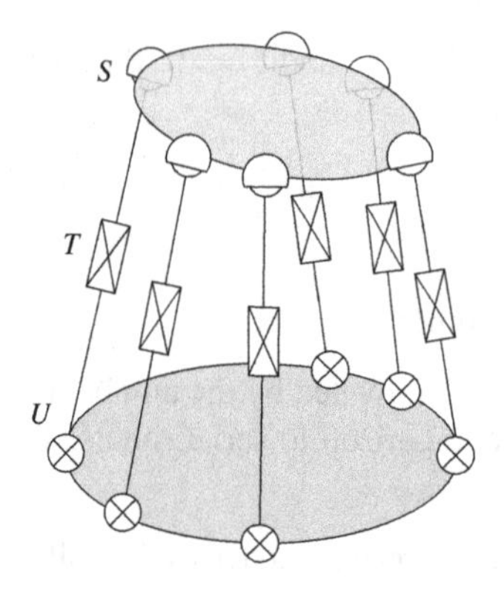

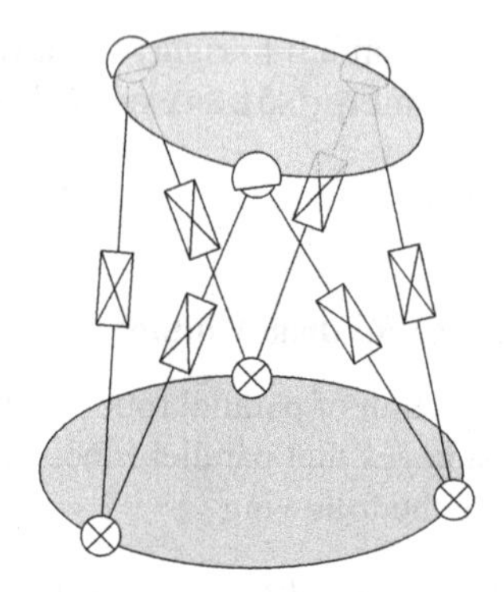

Fig. 6.4 The Stewart-Gough platform

Stewart-Gough platform

A general Stewart-Gough platform is shown on the left side of Fig. 6.4. According to our denomination, the mechanism is of type 6UTS. The robot contains $n = 18$ joints, $N = 13$ moving links and the sum of $f_i, i = 1, 2, \ldots n$ is 36. This gives the expected result

$$F = 6(13 - 18) + 36 = 6$$

degrees of freedom. The platform of this robot can be spatially positioned and oriented under the control of six motors, which are typically the six translations. By shortening or expanding the legs (changing the lengths of the legs) the platform can be moved into a desired pose (position and orientation). A special advantage of the Stewart-Gough platform with the UTS legs is that loads acting on the platform are transferred to each particular leg in the form of a longitudinal force in the direction of the leg and there is no transverse loading on the legs. This peculiarity allows excellent dynamic performances.

The number of degrees of freedom of a UTS leg is six and the number of constraints is zero. If we consider Grübler's formula (6.5) it is easy to verify that the number of UTS legs does not affect the number of degrees of freedom of the robot and that the mobility of the Stewart-Gough platform does not depend on the number of legs. A robot with only one UTS leg, which is a serial robot, possesses six degrees of freedom, the same as the fully parallel original six-legs Stewart-Gough robot.

The six-legged mechanism on the right side of Fig. 6.4 schematically represents the original Stewart-Gough platform which has a central-symmetrical star shape. In this arrangement, two by two legs are clamped in one point in which two overlapping coincident spherical (or universal) joints are placed. Therefore, the number of independent spherical joints is six and the same is the number of universal joints. The overlapping joints not only simplify the construction but also allow easier computations of the robot kinematics and dynamics.

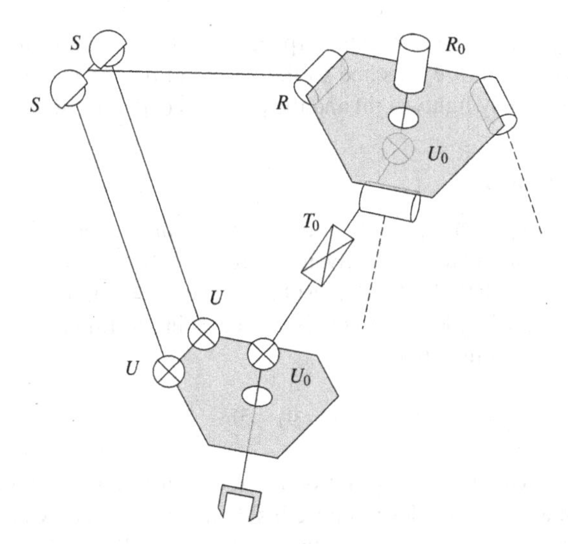

Fig. 6.5 The Delta robot

Delta robot

Due to its specific motion characteristics and its numerous applications in industry, the Delta robot found its place among robot manufacturers (see Fig. 6.5). The kinematics of this robot is very sophisticated. The main objective of its creator was to create a lightweight robot with extreme dynamic performances.

The fixed base of the robot is the upper hexagon while the lower hexagon represents the moving platform. The robot has three lateral legs. Only one is presented in the figure, with one R joint, two S joints and two U joints; the other two legs are symbolically drawn with a dotted line. There is also an independent middle leg $R_0U_0T_0U_0$ which has no influence on the motion of the platform. There is a parallelogram mechanism between the middle of the leg and the base, which consists of two spherical joints S and two universal joints U. Each leg, therefore, has 3 links and 5 joints. Without considering the middle leg, the number of degrees of freedom of the mechanism is

$$F = 6(10 - 15) + 33 = 3.$$

The pose of the platform is determined by only three variables. In the original version of the Delta robot the three rotation angles R in lateral legs are controlled by motors. Due to the parallelogram structure of the legs, the platform executes only translation and is always parallel to the base.

The purpose of the middle leg is to transfer the rotation R_0 across the platform to the gripper at the end-point of the robot. It acts as a telescoping driveshaft for rotating the gripper. This leg is a cardan joint with two universal joints U_0 separated by a translational joint T_0. In all, the mechanism has four degrees of freedom: three translational, enabling the spatial position of the gripper and one rotational, enabling rotation of the gripper about an axis perpendicular to the platform. All actuators of the Delta mechanism are attached to the base and do not move. Therefore the mechanism is extremely lightweight and the platform can move with high velocities and accelerations.

Planar parallel robots

The following examples are planar parallel robots which operate in a given plane where $\lambda = 3$. The first example is given in Fig. 6.6 left. The robot contains three legs of the type RTR-RRR-RRR. As a result we have $N = 7$ and $n = 9$ and the total number of degrees of freedom in joints is 9. According to Eq. (6.3), the number of degrees of freedom of this robot is

$$F = 3(7 - 9) + 9 = 3.$$

The result is expected since all legs introduce zero constraints (6.5). Consequently, the platform can achieve any desired pose inside the workspace. Note that in plane two degrees of freedom are needed for the position (translations in x–y plane) and one degree of freedom for the orientation (a rotation about z axis). To activate this robot three motors are needed. To attach the motors, we can select any of the nine joints. Usually we prefer the joints attached to the base so that the motors are not moving and their weight does not influence the robot dynamics. In a specific case, the translational joint can also be motorized using an electric spindle or a hydraulic cylinder.

In Fig. 6.6 right a similar planar parallel robot is presented, its structure is RTR-RR-RR. Here, we can see that each of the two RR legs introduce one constraint. According to Eq. (6.5), the number of degrees of freedom of this parallel robot is

$$F = 3 - 2 = 1.$$

The robot is controlled using one motor. The platform has limited mobility and can only move along a curve in plane x–y. We can, for example, either position the

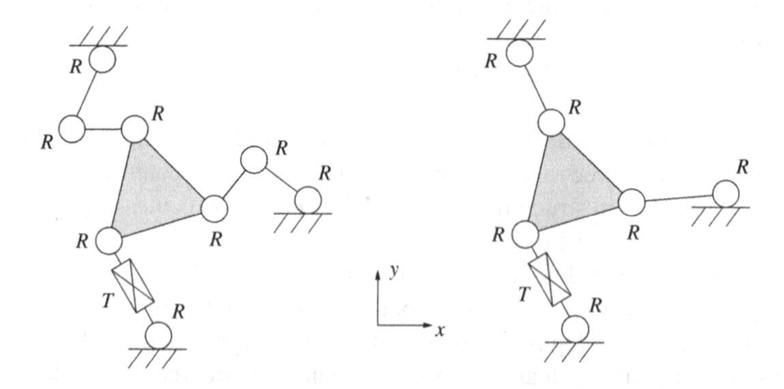

Fig. 6.6 Planar parallel robots

platform along x axis without having control over y and the platform's orientation or, alternatively, orient the platform without having control over its position in x and y.

Parallel humanoid shoulder

Parallel mechanisms are very common in nature, in the human body or in animals. It is, therefore, no surprise that the models of parallel robots can be efficiently used in simulating biomechanical properties of humans where muscles and ligaments stretched over the joints form various parallel kinematic structures. For instance, the shoulder complex can be represented by two basic compositions, the so-called inner joint, which includes the motion of the clavicle and the scapula with respect to the trunk, and the so-called outer joint, which is associated with the glenohumeral joint. In today's humanoid robotics, the motion of the inner joint is typically neglected because of its mechanical complexity. Nevertheless, its contribution to human motion, reachability of the arm and dynamics is crucial.

A parallel shoulder mechanism representing the inner shoulder was proposed in the literature. Its motion is shown in Fig. 6.7. The proposed structure is TS-3UTS. There is a central leg T_0S_0 with four degrees of freedom and two constraints. Around the axis of the central leg three UTS lateral legs are attached possessing six degrees of freedom each, their number of constraints is zero. According to Eq. (6.5), the number of degrees of freedom of the robot is

$$F = 6 - 2 = 4.$$

The robot can produce a complete orientation of the platform (about three principal orientation angles), and can expand or shrink similarly to the human shoulder. The arm is attached to this platform through the glenohumeral joint. The inner shoulder joints, as it is proposed, precisely mimic the motion of the arm, including shrugging

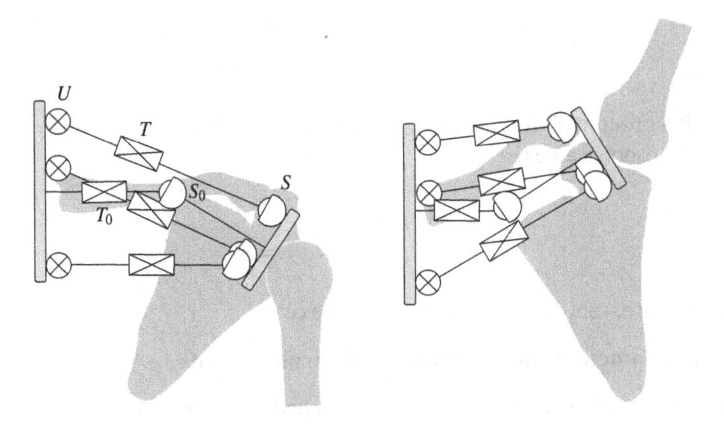

Fig. 6.7 Parallel robot mimicking the inner shoulder mechanism

and avoiding collisions with the body, and provides excellent static load capacity and dynamic capabilities.

6.3 Modelling and Design of Parallel Robots

The majority of parallel robots which appear in industry or in research laboratories possess symmetrical kinematic arrangements. From the point of view of their construction, it is useful that they are composed of the same mechanical elements. Symmetry also contributes to making their mathematical treatment simpler.

One common group of kinematic arrangements is represented by the previously described shoulder robot. This group contains a central leg with ν_1 degrees of freedom around which there are symmetrically placed lateral legs, which are often of type UTS possessing $\nu_2, \nu_3, ..., \nu_K = \lambda$ degrees of freedom (and zero constraints). The central leg is therefore crucial to determine the kinematic properties of the whole robot, as the number of degrees of freedom of the robot is $F = \nu_1$.

The second group of kinematic arrangements are represented by the Stewart-Gough platform in which all the legs are identical and are usually of type UTS so that $\nu_1, \nu_2, ..., \nu_K = \lambda$. When $\nu_1, \nu_2, ..., \nu_K < \lambda$ only a small number of such robots are movable, most of their structures are with zero or negative degrees of freedom. Robots with a negative number of degrees of freedom are referred to as overconstrained.

Consider the second group of robots (Gough-Stewart-like kinematic structure) with a single motor in each leg. Such a robot must have $K = F$ legs, as a robot with $K < F$ cannot be controlled. It is easy to verify that only the following robots can exist in space (where $\lambda = 6$)

$$\begin{aligned} &K = 1, \ \nu_1 = 1 \\ &K = 2, \ \nu_1 = \nu_2 = 4 \\ &K = 3, \ \nu_1 = \nu_2 = \nu_3 = 5 \\ &K = 6, \ \nu_1 = \nu_2 = ... = \nu_6 = 6 \end{aligned}$$

Robots in this group with four and five legs do not exist. In plane, where $\lambda = 3$, only the following robots can exist

$$\begin{aligned} &K = 1, \ \nu_1 = 1 \\ &K = 3, \ \nu_1 = \nu_2 = \nu_3 = 3 \end{aligned}$$

In the planar case, robots with two legs do not exist.

Kinematic parameters and coordinates of parallel robots

In Fig. 6.8 the coordinate frame x–y–z is attached to the moving platform, while x_0–y_0–z_0 is fixed to the base. The position of the platform is given with respect to the fixed coordinate frame by vector $\mathbf{r}$; its components are r_x, r_y, r_z. The orientation

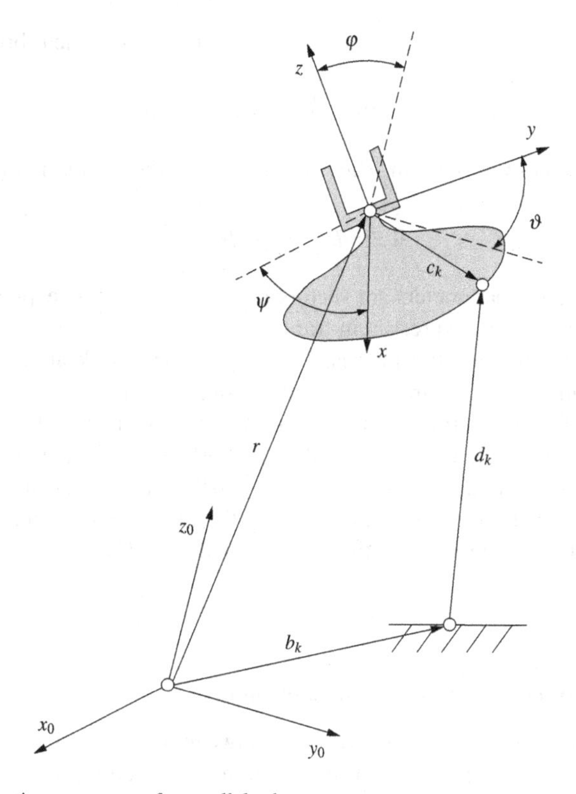

Fig. 6.8 Kinematic parameters of a parallel robot

of the platform can be described by a chosen triplet of orientation angles ψ, ϑ, φ occurring between both coordinate frames (see Chap. 4 for details).

Vector $\mathbf{b}_k$ defines the attachment of leg k to the base expressed in frame x_0–y_0–z_0, while vector $\mathbf{c}_k$ defines the attachment of the same leg to the platform in frame x–y–z. The vectors

$$\mathbf{d}_k = \mathbf{r} + \mathbf{R}\mathbf{c}_k - \mathbf{b}_k,\ k = 1, 2, ..., K, \tag{6.6}$$

describe the geometry of the robot legs expressed in coordinate frame x_0–y_0–z_0. Here, $\mathbf{R} = \mathbf{R}(\psi, \vartheta, \varphi)$ is the 3×3 rotation matrix which transforms the coordinate frame x–y–z into x_0–y_0–z_0. Equation (6.6) can also be formulated in a homogeneous form as follows

$$\mathbf{d}_k = \mathbf{H}\mathbf{c}_k,\ k = 1, 2, \ldots, K, \tag{6.7}$$

where the homogeneous transformation matrix is

$$\mathbf{H} = \begin{bmatrix} & \mathbf{R} & & \mathbf{r} - \mathbf{b}_k \\ 0 & 0 & 0 & 1 \end{bmatrix}. \tag{6.8}$$

We assume that the leg lengths are the joint coordinates of the robot

$$q_k = \|\mathbf{d}_k\|, \ k = 1, 2, \ldots, K, \tag{6.9}$$

where $\| \cdot \|$ indicates vector norm. They are elements of the vector of joint coordinates

$$\mathbf{q} = (q_1, q_2, ..., q_K)^{\mathrm{T}}.$$

The robot kinematic parameters are vectors $\mathbf{b}_k$, $k = 1, 2, \ldots, K$ expressed in frame x_0–y_0–z_0 and vectors $\mathbf{c}_k$ expressed in frame x–y–z.

Once we have defined the internal coordinates, let's look at what the robot's external coordinates are. In parallel robots they usually represent some characteristics in the motion of the platform to which the end-effector is attached. In most cases, the chosen external coordinates are the position and orientation of the platform, the so-called Cartesian coordinates. In space where $\lambda = 6$ they include the three components r_x, r_y, r_z of the position vector in Fig. 6.8, and the three orientation angles ψ, ϑ, φ, so that the vector of external coordinates is defined as follows

$$\mathbf{p} = (r_x, r_y, r_z, \psi, \vartheta, \varphi)^{\mathrm{T}}.$$

Inverse and direct kinematics of parallel robots

From the control point of view, the relation between the external and internal coordinates is of utmost importance. Their relationship is, similarly to serial robots, determined by very complicated algebraic trigonometric equations.

The inverse kinematics problem of parallel robots requires determining the internal coordinates $\mathbf{q}$, which are the leg lengths, from a given set of external coordinates $\mathbf{p}$, which represent the position and orientation of the platform. For a given set of external coordinates $\mathbf{p}$ the internal coordinates can be obtained by simply solving Eq. (6.7). Here, unlike in serial robots, it is important to recognize that the values of the external coordinates uniquely define the leg lengths of the parallel robot and the computation is straightforward.

The direct kinematics problem of parallel robots requires determining the external coordinates $\mathbf{p}$ from a given set of joint coordinates $\mathbf{q}$ (Fig. 6.9). This problem is extremely complicated in mathematical terms and the computation procedures are cumbersome. In general, it is not possible to express the external coordinates as explicit functions of the internal coordinates, whereas with serial robots this is quite straightforward. Usually, these are coupled trigonometric and quadratic equations which can be solved in closed-form only in special cases. There exist no rules as how to approach symbolic solutions. The following difficulties are common:

Nonexistence of a real solution. For some values of internal coordinates real solutions for the external coordinates do not exist. Intervals of internal coordinates when this can happen cannot be foreseen in advance.

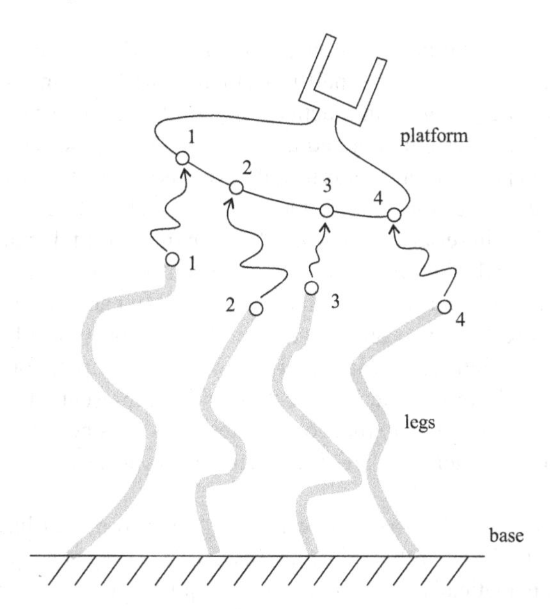

Fig. 6.9 The direct kinematics problem consists of finding the pose of the platform corresponding to the length of the legs. Leg end-points need to match corresponding points on the platform (e.g., 1 – 1)

Multiple solutions. For a given set of internal coordinates, there exist multiple solutions for the external coordinates. The number of solutions for a given combination of leg lengths depends on the kinematic structure of the mechanism. The general Stewart-Gough platform has forty possible solutions of the direct kinematics problem. For a selected combination of leg lengths there exist forty different poses of the platform. In addition, sometimes two poses of the platform cannot be transitioned between as the legs get entangled. In such cases, the platform could transit from one pose into another only by dismantling the legs in the first pose and reassembling them in the new pose.

Nonexistence of closed-form solutions. Generally for a given set of joint coordinates, it is not possible to find an exact solution to the direct kinematics problem, even if a real solution exists. In such cases we use numerical techniques which may not necessarily converge and may not find all the solutions.

Design of parallel robots

The design of parallel robots depends on desired performance, flexibility, mobility, and load capacity as well as the actual workspace.

In considering the workspaces for both parallel and serial robots, we are referring to the reachable workspace and the dexterous workspace. One of the main drawbacks of parallel robots is their small workspace. The main goal in workspace analysis is, therefore, to determine if a desired trajectory lies inside the robot workspace. The size of the workspace in parallel robots is limited by the ranges in the displacements of the legs, displacements of passive joints, and, particularly, by interference between the legs of the robot. Even with small movements, the legs can collide with each other. The interlacing of legs is in practice a major obstacle in a robot's motion and its reachability. The determination and analysis of robot workspace is in general a tedious process. In parallel robots it is usually even more complex, depending on the number of degrees of freedom and the mechanism's architecture.

The effect of load in serial robots is usually seen in terms of dynamics, which to a large extent includes the inertia of the links. The contribution of an external force is typically smaller and in many cases can be neglected. In parallel robots with a large number of legs, where the links are very light and the motors typically attached to the fixed base, the robot statics plays an important role. The computation of robot statics is related to the well-known Jacobian matrix which represents the transformation between the external and the internal coordinates. This goes beyond the scope of our book, but considerable literature, articles, and textbooks are available to the interested reader.

In practice, we can often see a Stewart-Gough platform that has a structure as presented in Fig. 6.10. The robot contains (instead of six legs of type UTS) six legs of type STS. Kinematically, this architecture is quite unusual and redundant. The robot has too many degrees of freedom. Each leg possesses 7 degrees of freedom

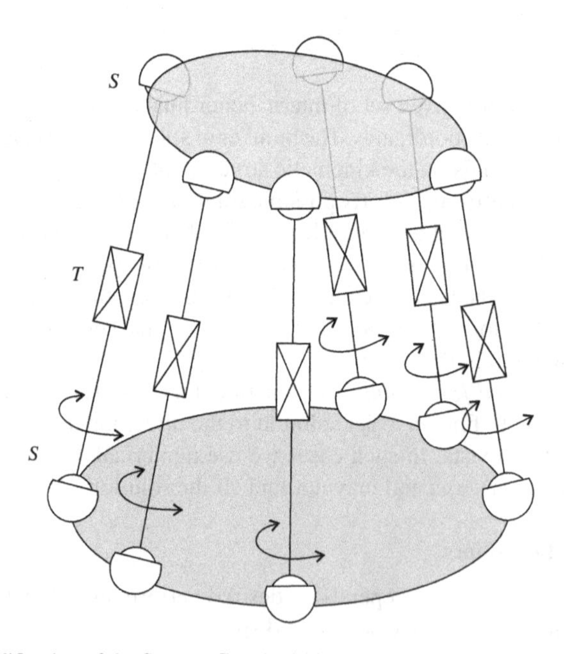

Fig. 6.10 A modification of the Stewart-Gough platform

which corresponds to -1 constraints. According to Grübler's formula (6.5), the number of degrees of freedom of the robot is

$$F = 6 - (-6) = 12.$$

It is important to note that six of the twelve degrees of freedom are manifested as rotations of the legs around their own axes. These rotations have no influence on the movement of the platform. Thus, the robot can still be motorized by only six motors that change the length of the legs, affecting the translation T, while the rotations around the leg axes can be left passive and can freely change. The advantages of this construction are that S joints are easier to build than U joints (and therefore cheaper), and that the passive rotations around the leg axes enable more flexibility when connecting power and signal cables, as these are often arranged along the legs from the base to the robot platform.

Chapter 7
Robot Sensors

The human sensory system encompasses sensors of vision and hearing, kinesthetic sensors (movement, force, and touch), sensors of taste and smell. These sensors deliver input signals to the brain which uses this sensory information to build its own image of the environment and takes decisions for further actions. Similar requirements are valid also for robot mechanisms. However, because of the complexity of human sensing, robot sensing is limited to fewer sensors.

The use of sensors is of crucial importance for efficient and accurate robot operation. Robot sensors can be generally divided into: (1) proprioceptive sensors assessing the internal states of the robot mechanism (positions, velocities, and torques in the robot joints); and (2) exteroceptive sensors delivering to the controller the information about the robot environment (force, tactile, proximity and distance sensors, robot vision).

7.1 Principles of Sensing

In general, sensors convert the measured physical variable into an electrical signal which can be in a digital form assessed by the computer. In robotics we are predominantly interested in the following variables: position, velocity, force, and torque. By the use of special transducers these variables can be converted into electrical signals, such as voltage, current, resistance, capacity, or inductivity. Based on the principle of conversion the sensors can be divided as follows:

- Electrical sensors—the physical variable is directly transformed into an electrical signal; such sensors are for example potentiometers or strain gauges;
- Electromagnetic sensors—use the magnetic field for the purposes of physical variable conversion; an example is the tachometer;

M. Mihelj et al., *Robotics*, https://doi.org/10.1007/978-3-319-72911-4_7

- Optical sensors—use light when converting the signals; an example of such a sensor is the optical encoder.

7.2 Sensors of Movement

Typical sensors of robot movements are potentiometers, optical encoders, and tachometers. They all measure the robot movements inside the robot joint. Where in the joint to place the movement sensor is important, as well as how to measure the motion parameters.

7.2.1 Placing of Sensors

Let us first consider a sensor of angular displacement. It is our aim to measure the angle in a robot joint which is actuated by a motor through a reducer with the reduction ratio k_r. Using a reducer we decrease the joint angular velocity by the factor k_r with respect to the angular velocity of the motor. At the same time the joint torque is increased by the same factor. It is important whether the movement sensor is placed before or after the reducer. The choice depends on the task requirements and the sensor used. In an ideal case we mount the sensor before the reducer (on the side of the motor), as shown in Fig. 7.1. In this way we measure directly the rotations of the motor. The sensor output must be then divided by the reduction ratio, to obtain the joint angle.

Let us denote by ϑ the angular position of the joint, ϑ_m as the angular position of the corresponding motor, and k_r the reduction ratio of the reducer. When the sensor

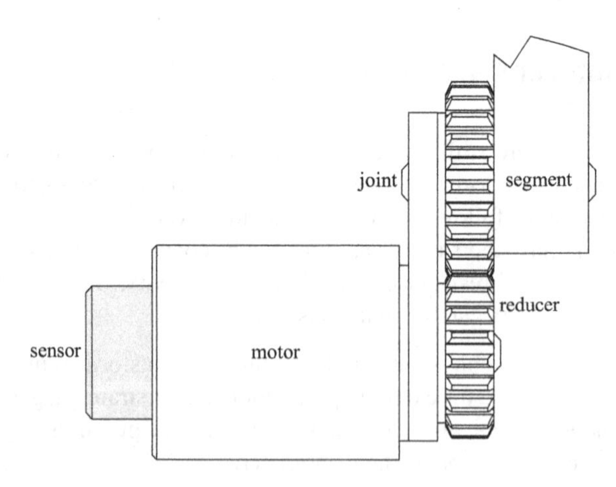

Fig. 7.1 Mounting of the sensor of movement before the reducer

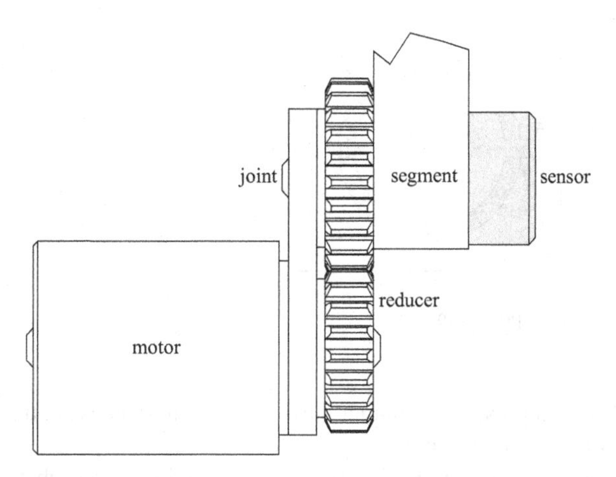

Fig. 7.2 Mounting of the sensor of movement behind the reducer

is placed before the reducer, its output is equal to the angle ϑ_m. The variable which we need for control purposes is the joint angle ϑ, which is determined by the ratio

$$\vartheta = \frac{\vartheta_m}{k_r}. \tag{7.1}$$

By differentiating the Eq. (7.1) with respect to ϑ_m we have

$$\frac{d\vartheta}{d\vartheta_m} = \frac{1}{k_r} \quad \text{thus} \quad d\vartheta = \frac{1}{k_r} d\vartheta_m, \tag{7.2}$$

which means that the sensor measurement error is reduced by the factor k_r. The advantage of the placement of the sensor before the reducer is in getting more accurate information about the joint angular position.

Another sensor mounting possibility is shown in Fig. 7.2. Here, the sensor is mounted behind the reducer. In this way the movements of the joint are measured directly. The quality of the control signal is decreased, as the sensor measurement error (which is now not reduced) directly enters the joint control loop. As the range of motion of the joint is by the factor k_r smaller than that of the motor, sensors with smaller range of motion can be used. Sometimes we cannot avoid mounting the motion sensor into the joint axis. It is important, therefore, that we are aware of the deficiency of such a placement.

7.2.2 *Potentiometer*

Figure 7.3 presents a model of a rotary potentiometer and its components. The potentiometer consists of two parts: (1) resistive winding and (2) movable wiper.

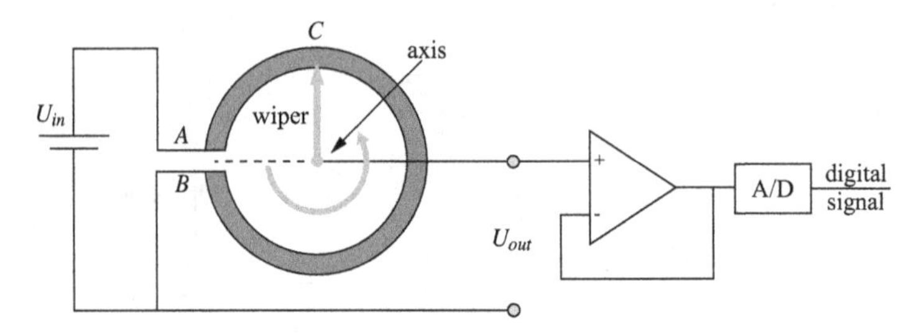

Fig. 7.3 The model of a potentiometer

The potentiometer represents a contact measuring method, because the wiper slides along the circular resistive winding.

Potentiometers are generally placed behind the reducer in such a way that the potentiometer axis is coupled to the joint axis. Let us suppose that point B represents the reference position of the potentiometer belonging to the joint. The resistance of the potentiometer along the winding $\widehat{AB}$ equals R, while r represents the resistance of the $\widehat{CB}$ part of the winding. The angle of the wiper with respect to the reference position B is denoted by ϑ (in radians). When the resistance along the circular winding of the potentiometer is uniform and the distance between the points A and B is negligible, we have the following equation

$$\frac{r}{R} = \frac{\widehat{CB}}{\widehat{AB}} = \frac{\vartheta}{2\pi}. \tag{7.3}$$

Let us suppose that the potentiometer is supplied by the voltage U_{in}. The output voltage measured on the wiper is equal to

$$\frac{U_{out}}{U_{in}} = \frac{r}{R} = \frac{\vartheta}{2\pi}, \tag{7.4}$$

or

$$U_{out} = \frac{U_{in}}{2\pi}\vartheta. \tag{7.5}$$

By measuring the output voltage U_{out}, the angular position ϑ is determined.

7.2.3 *Optical Encoder*

The contact measurement approach to the robot joint angle using potentiometers has several deficiencies. The most important is the relatively short lifespan because of high wear and tear. In addition, the most adequate placement is directly in the joint axis (behind the reducer) and not on the motor axis (before the reducer). The

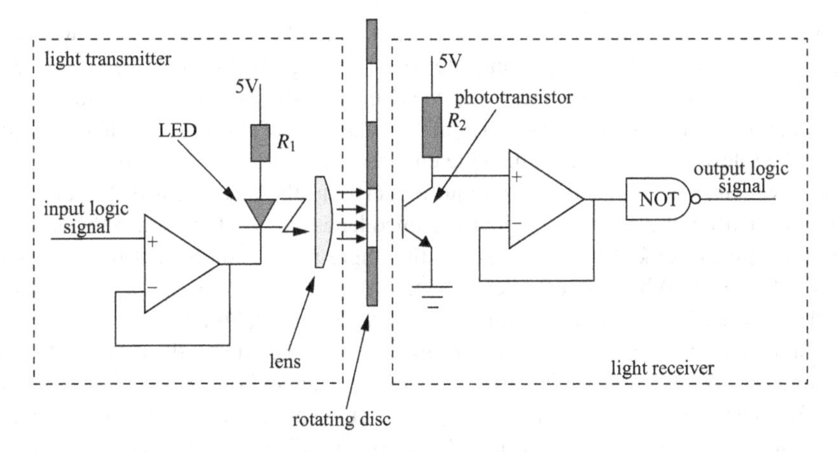

Fig. 7.4 The model of optical encoder

most widely used movement sensors in robotics are therefore optical encoders which provide contact-less measurement.

The optical encoder is based on the transformation of the joint movement into a series of light pulses, which are further converted into electrical pulses. To generate the light pulses, a light source is needed, usually a light emitting diode. The conversion of light into electrical pulses is performed by a phototransistor or a photodiode converting light into electrical current.

The model of an optical encoder assessing the joint angular position is presented in Fig. 7.4. It consists of a light source with lens, light detector, and a rotating disc with slots, which is connected to either motor or joint axis. On the rotating disc there is a track of slots and interspaces, which alternately either transfer or block the light from the light emitting diode to the phototransistor. The logical output of the sensor is high when the light goes through the slot and hits the phototransistor on the other side of the rotating plate. When the path between the light emitting diode and the phototransistor is blocked by the interspace between two slots, the logical output is low.

The optical encoders are divided into absolute and incremental. In the further text we shall learn about their most important properties.

7.2.3.1 Absolute Encoder

The absolute optical encoder is a device which measures the absolute angular position of a joint. Its output is a digital signal. In a digital system each logical signal line represents one bit of information. When connecting all these bits into a single logical state variable, the number of all possible logical states determines the number of all absolute angular positions that can be measured by the encoder.

Let us suppose that we wish to measure the angular rotation of 360° with the resolution of 0.1°. The absolute encoder must distinguish between 3600 different logical states, which means that we need at least 12 bits to assess the joint angles with the required resolution. With 12 bits we can represent 4096 logical states. An important design parameter of absolute encoders is therefore the number of logical states, which depends on the task requirements and the placement of the encoder (before or after the reducer). When the encoder is placed before a reducer with the reduction ratio k_r, the resolution of the angle measurement will be increased by the factor k_r. When the encoder is behind the reducer, the necessary resolution of the encoder is directly determined by the required resolution of the joint angle measurement. All logical states must be uniformly engraved into the rotating disc of the encoder. An example of absolute encoder with sixteen logical states is shown in Fig. 7.5. The sixteen logical states can be represented by four bits. All sixteen logical states are engraved into the surface of the rotating disc. The disc is in the radial direction divided into four tracks representing the four bits. Each track is divided into sixteen segments corresponding to the logical states. As the information about the angular displacement is represented by four bits, we need four pairs of light emitting diodes and phototransistors (one pair for each bit). With the rotation of the disc, which is connected to either motor or joint axis, the output signal will change according to the logical states defined by the order of segments (Grey code, where two successive values differ in only one bit, is typically used in absolute encoders).

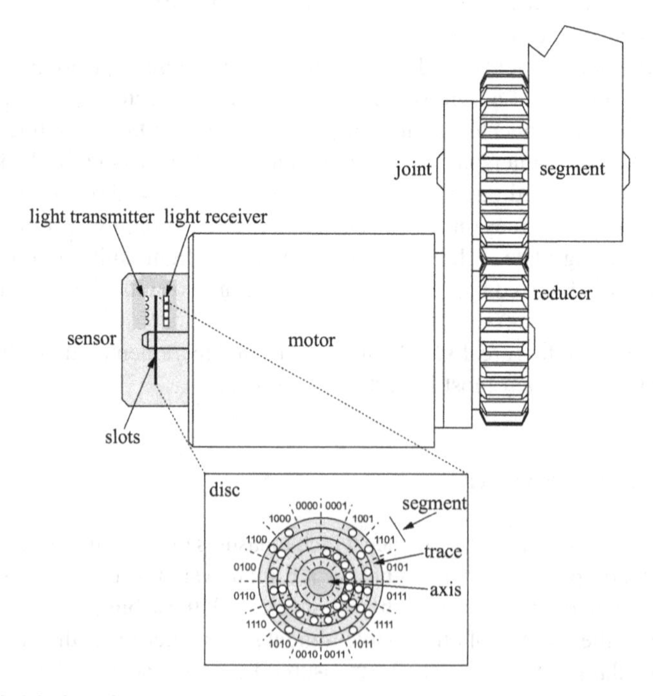

Fig. 7.5 Model of absolute encoder

The absolute encoder does not determine only the angular position of the joint but also the direction of rotation.

7.2.3.2 Incremental Encoder

In contrast to absolute encoders, the incremental encoders only supply information about the changes in angular joint position. The advantages of incremental encoders, compared to the absolute encoders, are their simplicity, smaller dimensions, and (most importantly) low cost. This can be achieved by lowering the number of the tracks on the rotating disc to only a single track. Instead of having as many tracks as the number of the bits necessary for the representation of all required logical states, we have now only one track with even graduation of the slots along the rim of the disc. Figure 7.6 shows a model of an incremental encoder. A single track only requires a single pair of light emitting diode and phototransistor (optical pair). During rotation of the encoded disc a series of electrical pulses is generated. The measurement of the joint displacement is based on counting of these pulses. Their number is proportional to the robot joint displacement. The incremental encoder shown in Fig. 7.6 generates eight pulses during each rotation. The resolution of this encoder is

$$\Delta\vartheta = \frac{2\pi}{8} = \frac{\pi}{4}. \tag{7.6}$$

By increasing the number of the slots on the disc, the resolution of the encoder is increased. By denoting the number of the slots as n_c, a general equation for the encoder resolution can be written

$$\Delta\vartheta = \frac{2\pi}{n_c}. \tag{7.7}$$

The encoder with one single track is only capable of assessing the change in the joint angular position. It cannot provide information about the direction of rotation or the absolute joint position. If we wish to apply the incremental encoders in robot control, we must determine: (1) the home position representing the reference for the measurement of the change in the joint position and (2) the direction of rotation.

The problem of the home position is solved by adding an additional reference slot on the disc. This reference slot is displaced radially with respect to the slotted track measuring the angular position. For detection of the home position, an additional optical pair is needed. When searching for the reference slot, the robot is programmed to move with low velocity, as long as the reference slot or the end position of the joint range of motion is reached. In the latter case the robot moves in the opposite direction towards the reference slot.

The problem of determining the direction of rotation is solved by another pair of light emitting diode and phototransistor. This additional optical pair is tangentially and radially displaced from the first optical pair as shown in Fig. 7.6. When the disc

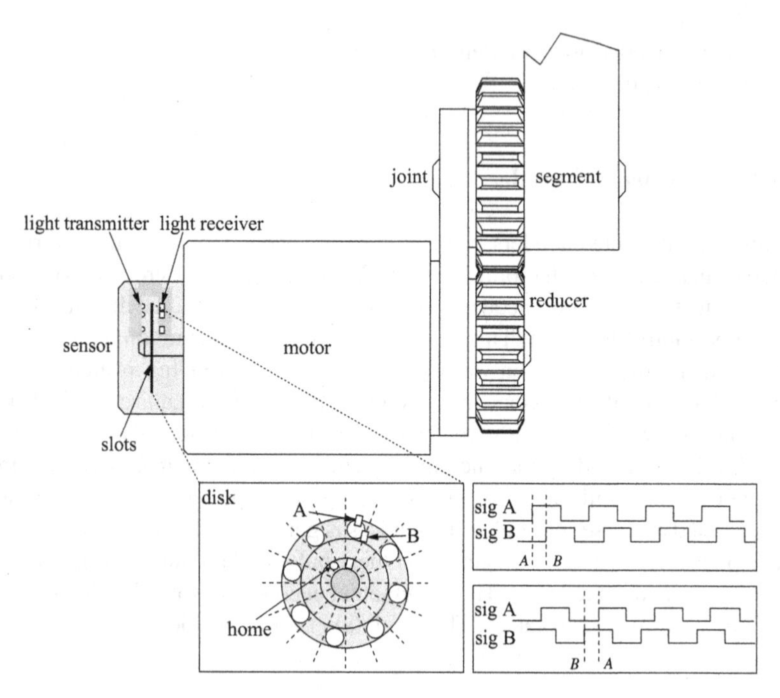

Fig. 7.6 Model of incremental encoder. The series of pulses for positive (above) and negative (below) direction of rotation

is rotating, two signals are obtained, which are, because of the displacement of the optical pairs, shifted in phase. This shift in phase occurs because each slot on the disc first reaches the first optical pair and after a short delay also the second pair. The optical components are usually placed in such a way that the phase shift of $\pi/2$ is obtained between the two signals. During the rotation in clockwise direction the signal B is phase-lagged for $\pi/2$ behind the signal A. During counter clockwise rotation the signal B is in phase-lead of $\pi/2$ with respect to the signal A (Fig. 7.6). The direction of the encoder rotation can be determined upon the basis of the phase shifts between signals A and B. Another advantage of having two optical pairs is the possibility of counting all the changes in both the A and B signals. The approach known as quadrature decoding enables measurement resolution of four-times the nominal encoder resolution.

7.2.4 Magnetic Encoder

In contrast to optical encoders the magnetic encoder uses magnetic field for measuring position. This can be achieved by using a series of magnetic poles (2 or more) on the sensor rotor to represent the encoder position to a magnetic sensor. The rotor turns

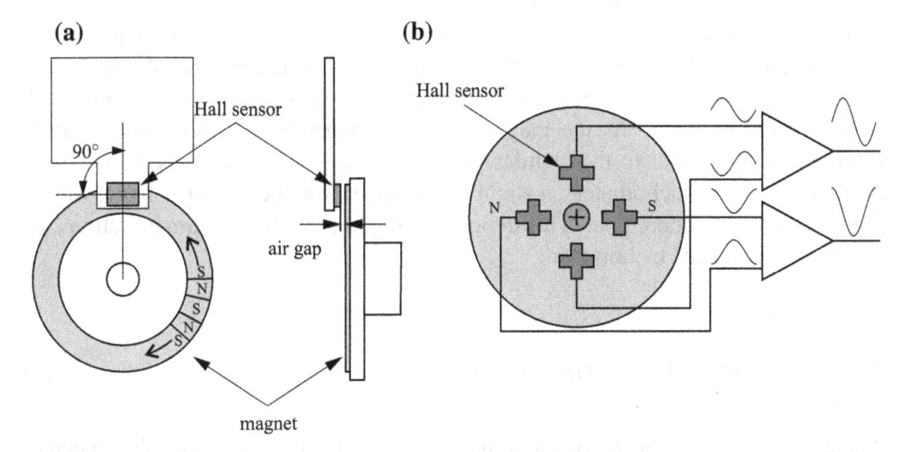

Fig. 7.7 Model of magnetic encoder: **a** Hall sensor and alternating north and south poles and **b** multiple Hall sensors with diametrically magnetized rotating magnet

with the shaft and contains alternating evenly spaced north and south poles around its circumference. The magnetic sensor (typically magneto-resistive or Hall effect) reads the magnetic pole positions. Hall sensors generate output voltage proportional to the strength of an applied magnetic field. Magneto-resistive sensors detect changes in resistance caused by a magnetic field. The principle of operation is shown in Fig. 7.7a.

Hall sensors can be used for angle measurement also when placed near a diametrically magnetized magnet that generates a sinusoidal waveform. The limitation of this method is the ambiguity at angles >90° in both directions from the zero crossing point. In order to extend the measurement range to 360°, refinement of the method is required. The problem can be solved by using multiple Hall sensors, rather than one, and placing them underneath a diametrically magnetized rotating magnet to generate multiple sinusoidal waveforms. Figure 7.7b shows four equally spaced Hall sensors generating four sinusoidal signals, each phase-shifted by 90° from its neighbor. Magnetic encoders are typically more robust than optical encoders.

7.2.5 Tachometer

The signal of the joint velocity can be obtained by numerical differentiation of the position signal. Nevertheless, direct measurement of the joint velocity with the help of a tachometer is often used in robotics. The reason is the noise introduced by numerical differentiation, which can greatly affect the quality of the robot control.

Tachometers can be divided into: (1) direct current (DC) and (2) alternate current (AC) tachometers. In robotics it is generally the simpler DC tachometers that are used. The working principle is based on a DC generator whose magnetic field is provided by permanent magnets. As the magnetic field is constant, the tachometer output voltage is proportional to the angular velocity of the rotor. Because commutators are used in the DC tachometers, a slight ripple appears in the output voltage, which cannot be entirely filtered out. This deficiency, together with other imperfections, is avoided by using AC tachometers.

7.2.6 Inertial Measurement Unit

Potentiometers and optical encoders measure joint displacements in robot mechanisms. When considering, for example, a robotic aerial vehicle, or a wheeled robot, these sensors do not provide information about the orientation of the device in space.

Measuring of object (robot) orientation in space is typically based on the magneto-inertial principle. This method combines a gyroscope (angular velocity sensor), accelerometer (linear acceleration sensor), and magnetometer (measures orientation relative to the Earth's magnetic field and is not regarded as an inertial sensor).

The method will be illustrated with the example of a rigid pendulum equipped with a two-axis accelerometer (measures accelerations along two perpendicular axes) and a single-axis gyroscope (Fig. 7.8). Both sensors give the measured quantities in their own coordinate frames, which are attached to the center of the sensor and have their axes parallel to the x and y axes of the coordinate frame attached to the pendulum. Figure 7.8a shows a stationary pendulum while Fig. 7.8b shows a swinging pendulum. We are interested in the orientation of the pendulum relative to the reference coordinate frame x_0–y_0–z_0. Since the pendulum is only swinging around z axis, we are only actually interested in angle φ.

We first analyze stationary conditions. Since the angular velocity of a stationary pendulum is equal to zero, the gyroscope's output is also zero and the gyroscope tells us nothing about the pendulum's orientation. However, we can see that the accelerometer still measures the gravitational acceleration. Since the accelerometer is at an angle of φ relative to the gravitational field, two acceleration components are measured: $\mathbf{a}_x$ and $\mathbf{a}_y$. The vector sum of both components gives the gravitational acceleration. Figure 7.8a shows that the angle between vectors $\mathbf{g}$ and $\mathbf{a}_y$ is equal to φ. Since the scalar values of accelerations a_x and a_y are known, we can now determine the pendulum angle

$$\varphi = \arctan \frac{a_x}{a_y}. \tag{7.8}$$

The accelerometer thus allows the pendulum's angle to be measured in stationary conditions. For this reason, accelerometers are frequently used as inclinometers.

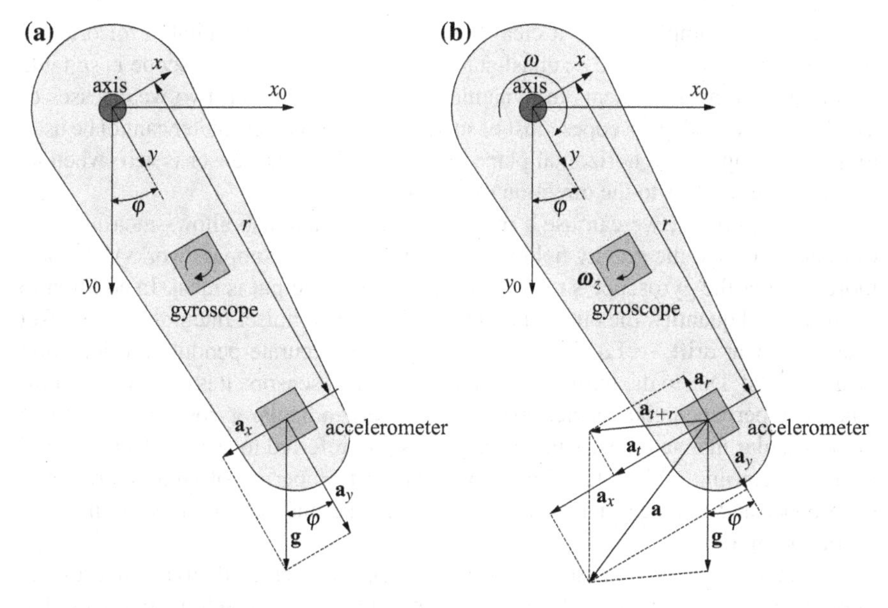

Fig. 7.8 Example of using an inertial measurement system to measure the angle of a pendulum: left figure—stationary pendulum, right figure—swinging pendulum

The conditions in a swinging pendulum are quite different. Since swinging is an accelerated rotational movement, the accelerometer is affected not only by the gravitational acceleration $\mathbf{g}$, but also by centripetal acceleration

$$\mathbf{a}_r = \boldsymbol{\omega} \times (\boldsymbol{\omega} \times \mathbf{r}) \tag{7.9}$$

and tangential acceleration

$$\mathbf{a}_t = \dot{\boldsymbol{\omega}} \times \mathbf{r}. \tag{7.10}$$

The total acceleration acting on the accelerometer is thus

$$\mathbf{a} = \mathbf{g} + \mathbf{a}_r + \mathbf{a}_t. \tag{7.11}$$

The equation used to calculate angle in stationary conditions (7.8) is no longer valid, therefore, the accelerometer cannot be used to calculate the angle of a swinging pendulum. However, the output of the gyroscope, which measures the angular velocity of the pendulum, is now also available. Since the angle of the pendulum can be calculated as the temporal integral of angular velocity, the following relation can be stated

$$\varphi = \varphi_0 + \int \omega dt, \tag{7.12}$$

where the initial orientation of the pendulum φ_0 must be known.

The given example makes it clear that an accelerometer is suitable for orientation measurements in static or quasi-static conditions while a gyroscope is suitable for orientation measurements in dynamic conditions. However, two weaknesses of accelerometers and gyroscopes must be mentioned. An accelerometer cannot be used to measure angles in a horizontal plane, as the output of the sensor is zero when its axis is perpendicular to the direction of gravity.

For this purpose, we can use a magnetometer, which also allows measurement of rotation around the gravity field vector (think of how a compass works). Furthermore, neither the gyroscope's nor the accelerometer's output is ideal. In addition to the measured quantity, the output includes an offset and noise. Integrating the offset causes a linear drift, so Eq. (7.12) does not give an accurate pendulum orientation measurement. Due to the weaknesses of the individual sensors, it is common to combine three perpendicular accelerometers, three perpendicular gyroscopes, and three perpendicular magnetometers into a single system, referred to as a magneto-inertial measurement unit (IMU). Combination of the best properties of an accelerometer, gyroscope, and magnetometer can give an accurate and reliable measurement of spatial orientation.

The angular velocity measured by the gyroscope is integrated, giving an estimate of orientation. Measurements from the accelerometer and magnetometer are used to directly calculate the sensor orientation with reference to the gravity and magnetic field vectors. This is achieved through sensor fusion, which can be done by using the Kalman filter.

7.3 Contact Sensors

The sensors considered so far provide information about robot pose and motion. They enable closing of the position and velocity control loop. In some robot tasks contact of the end-effector with the environment is required. Typical contact sensors used in robotics are tactile sensors and force and torque sensors. Tactile sensors measure parameters that define the contact between the sensor and an object.

Sensing consists in measurement of a point contact force and the spatial distribution of forces perpendicular to an area. By contrast, force and torque sensors measure the total forces being applied to an object.

7.3.1 Tactile Sensor

Robots can collect information about the environment also through touch. In order to increase robot manipulation capabilities, tactile sensors can be used in robotic fingers as shown in Fig. 7.9a. The sensor provides data about contact force distribution between the finger and the manipulated object. To increase robot safety (e.g., when

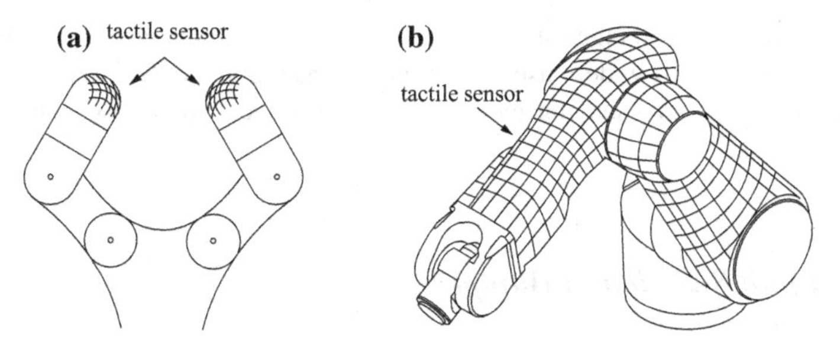

Fig. 7.9 Tactile sensor used in robot finger (left) and as robot skin (right)

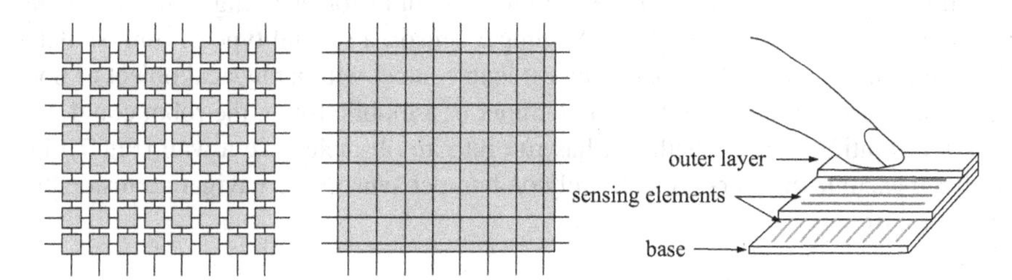

Fig. 7.10 A model of a tactile sensor

working with a human), tactile sensors can be used as an artificial robot skin that enables the robot to sense contacts with objects in the environment (Fig. 7.9b).

Tactile sensing is based on an array of touch sensors as shown in Fig. 7.10. The following sensing principles can be implemented in the array:

- deformation-based sensors—material surface deforms (changes length), when it is subjected to external forces; deformation is converted to electrical signals with strain gauges connected in a Wheatstone bridge;
- resistive sensors—electrical resistance changes with pressure of a material placed between two electrodes;
- capacitive sensors—sensing element is a capacitor whose capacitance changes with the applied force; force can produce either a change in the distance between capacitor plates or its area;
- optical sensors—sensing is typically based on light intensity measurement; intensity of light can be modulated by moving an obstruction or a reflective surface into the light path; the intensity of the received light is a function of displacement and hence of the applied force;
- piezoelectric sensors—materials, like quartz, have piezoelectric properties and can thus be used for tactile sensing; piezoelectric transducers are not adequate for static force transduction; this problem can be overcome by vibrating the sensor and detecting the difference in the vibration frequency due to the applied force;

- magnetic sensors—changes of magnetic flux density or magnetic coupling between circuits are the most widely used principles in magnetic tactile sensing;
- mechanical sensors—sensing elements are mechanical micro-switches with on and off states.

7.3.2 Limit Switch and Bumper

Limit switches are often used to control robot mechanisms. They can sense a single position of a moving part and are therefore suitable for ensuring that movement doesn't exceed a predefined limit. A bumper sensor, a special type of limit switch, for instance, will tell the robot whether it is in contact with a physical object or not. If the sensor is mounted on the front bumper of a mobile robot, the robot could use this information to tell whether it has run into an obstacle, like a wall (Fig. 7.11). Robotic vacuum cleaners typically rely on bumper sensors for navigating inside the home environment.

7.3.3 Force and Torque Sensor

In the simplest case the force measurement enables disconnection of the robot when the contact force exceeds a predetermined safety limit. In a more sophisticated case we use force sensors for control of the force between the robot end-effector and the environment. It is therefore not difficult to realize that the force sensor is placed into the robot wrist and is therefore often called the wrist sensor.

Strain gauges are usually used for the force measurements. The strain gauge is attached to an elastic beam which is deformed under the stress caused by the applied force. The strain gauge therefore behaves as a variable resistor whose resistance changes proportionally to its deformation. The wrist sensor must not influence the

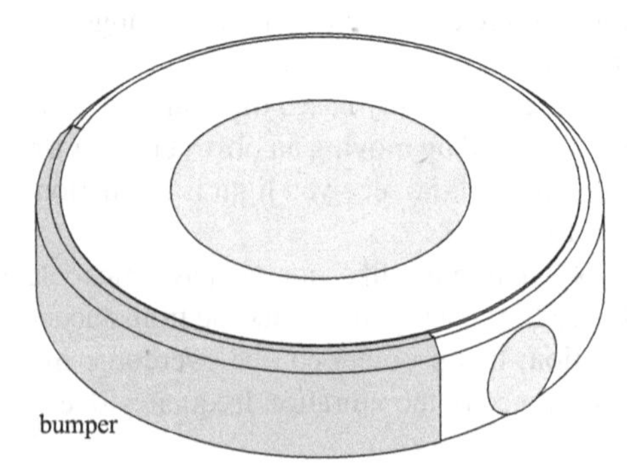

Fig. 7.11 Bumper sensors to be used on a mobile robot

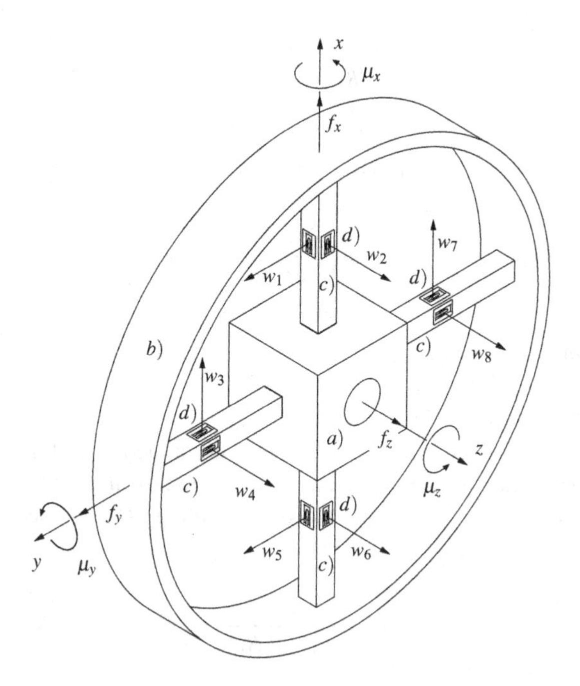

Fig. 7.12 Model of the force and torque sensor: **a** rigid body which is in contact with the robot end-effector, **b** rigid ring which is in contact with the robot environment, **c** elastic beams, and **d** strain gauge

interaction of the robot with the environment. This means that the wrist sensor must be sufficiently rigid. The robot wrist sensors are usually designed as shown in Fig. 7.12. The structure of the sensor is based on three components: (a) rigid inner part which is in contact with the robot end-effector; (b) rigid outer ring which is in contact with the robot environment; and (c) elastic beams interconnecting the outer and the inner ring. During contact of the robot with the environment, the beams are deformed by the external forces which causes a change in the resistance of the strain gauges.

The vector of the forces and torques acting at the robot end-effector is in the three-dimensional space represented by six elements, three forces and three torques.

The rectangular cross-section of a beam (shown in Fig. 7.12) enables the measurement of deformations in two directions. To measure the six elements of the force and torque vector, at least three beams, which are not collinear, are necessary. Four beams are used in the example in Fig. 7.12. There are two strain gauges attached to the perpendicular surfaces of each beam. Having eight strain gauges, there are eight variable resistances, R_1 to R_8. As the consequence of the external forces and torques, elastic deformations w_1 to w_8 occur resulting in changes in the resistances ΔR_1 to ΔR_8. The small changes in the resistance are, by the use of the Wheatstone bridge, converted into voltage signals (Fig. 7.13). To each of the eight variable

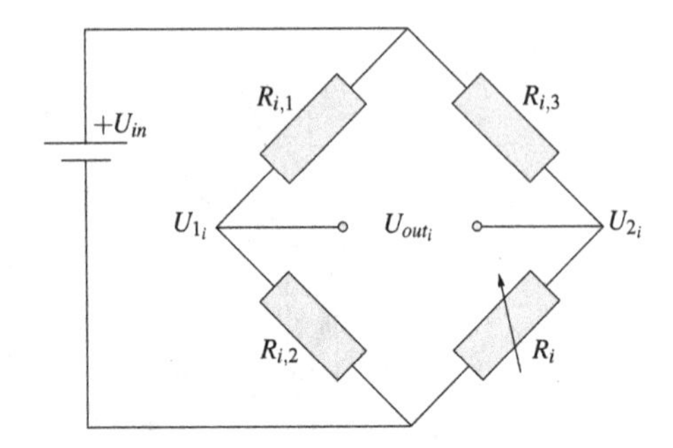

Fig. 7.13 The Wheatstone bridge

resistors $\{R_1 \ldots R_8\}$, three additional resistors are added. The three resistors are, together with the strain gauge, connected into the measuring bridge. The bridge is supplied with the U_{in} voltage, while the output voltage U_{out_i} is determined by the difference $U_{1_i} - U_{2_i}$. The U_{1_i} voltage is

$$U_{1_i} = \frac{R_{i,2}}{R_{i,1} + R_{i,2}} U_{in}, \tag{7.13}$$

while the U_{2_i} voltage is

$$U_{2_i} = \frac{R_i}{R_i + R_{i,3}} U_{in}. \tag{7.14}$$

The output voltage is equal to

$$U_{out_i} = \left(\frac{R_{i,2}}{R_{i,1} + R_{i,2}} - \frac{R_i}{R_i + R_{i,3}} \right) U_{in}. \tag{7.15}$$

By differentiating the Eq. (7.15) with respect to the variable R_i, the influence of the change of the strain gauge resistance on the output voltage can be determined

$$\Delta U_{out_i} = -\frac{R_{i,3} U_{in}}{(R_i + R_{i,3})^2} \Delta R_i. \tag{7.16}$$

Before application, the force sensor must be calibrated, which requires the determination of a 6×8 calibration matrix transforming the six output voltages into the three forces $\left[f_x \; f_y \; f_z\right]^T$ and three torques $\left[\mu_x \; \mu_y \; \mu_z\right]^T$

$$\left[f_x \; f_y \; f_z \; \mu_x \; \mu_y \; \mu_z\right]^T = \mathbf{K} \left[U_{out_1} \; U_{out_2} \; U_{out_3} \; U_{out_4} \; U_{out_5} \; U_{out_6} \; U_{out_7} \; U_{out_8}\right]^T, \tag{7.17}$$

where

$$\mathbf{K} = \begin{bmatrix} 0 & 0 & K_{13} & 0 & 0 & 0 & K_{17} & 0 \\ K_{21} & 0 & 0 & 0 & K_{25} & 0 & 0 & 0 \\ 0 & K_{32} & 0 & K_{34} & 0 & K_{36} & 0 & K_{38} \\ 0 & 0 & 0 & K_{44} & 0 & 0 & 0 & K_{48} \\ 0 & K_{52} & 0 & 0 & 0 & K_{56} & 0 & 0 \\ K_{61} & 0 & K_{63} & 0 & K_{65} & 0 & K_{67} & 0 \end{bmatrix} \tag{7.18}$$

is the calibration matrix with constant values K_{ij}.

7.3.4 Joint Torque Sensor

Often it is required or preferable to measure joint torques instead of robot end-effector forces. In such cases a joint torque sensor must be used. By measuring joint torques the robot can respond to forces applied anywhere on its mechanism. If the robot dynamic model is known, it is also possible to estimate end-effector forces.

As an example, consider Eq. (5.20). The inverse of this equation would give

$$\mathbf{f} = \mathbf{J}^{-T}(\mathbf{q})\tau. \tag{7.19}$$

It should be noted that the above equation would give exact end-effector forces only in static conditions and if gravity force does not affect joint torques. In other conditions, robot dynamic model (5.56) must be taken into account.

The principle of operation of torque sensor is typically similar to that of the wrist sensor. However, its mechanical structure is designed to fit onto the joint axis.

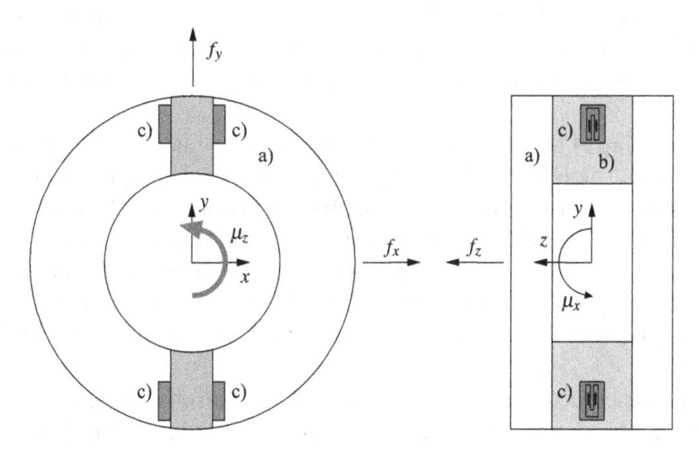

Fig. 7.14 Joint torque sensor structure measures torque μ_z: **a** frame, **b** elastic element, and **c** strain gauge

Thus, the sensor is integrated between the actuator (and possibly gear) and the robot segment. The sensor needs to guarantee high sensitivity to torsion, low sensitivity to non-torsional components, and high stiffness in all axes of forces and moment. Deformation of mechanical structure due to joint torque is measured by using strain gauges. A schematic representation of joint torque sensor is shown in Fig. 7.14.

7.4 Proximity and Ranging Sensors

Proximity and ranging sensors detect the presence of nearby objects without any physical contact. Consecutively they enable distinguishing between obstacles of different shapes and sizes as well as more efficient obstacle avoidance than contact sensors. Different methods can be used to detect obstacles from a distance. Methods based on magnetic and capacitive principles typically enable detecting proximity of an object but not its distance. When distance is relevant, active methods such as ultrasonic rangefinder, laser rangefinder, and infrared proximity sensor as well as passive methods based on cameras can be considered. All methods are characterized by high reliability and long operational life as they operate without physical contact between the sensor and the sensed object.

7.4.1 Ultrasonic Rangefinder

An ultrasonic rangefinder measures the distance to an object by using sound waves. Distance is measured by sending out a sound wave at an ultrasonic frequency (higher frequencies are better for short range and high precision needs) and listening for that sound wave to bounce back (Fig. 7.15a). The elapsed time between the sound wave being generated and the sound wave bouncing back, is used to calculate the distance between the sensor and the object (considering that the speed of sound in the air is approximately 343 m/s).

Understanding of the detection zone is important for successful object detection and avoidance. The beam width of ultrasonic rangefinder is typically described as a cone of a certain angle. This angle describes the arc at which the ultrasonic wave emanates from the transducer. However, at a certain distance the rate of expansion starts to decay as shown in Fig. 7.15b. An extension of the measurement area of an ultrasonic rangefinder can be achieved by using multiple sensor units facing at different angles (Fig. 7.15c). The problem of cross-talk needs to be considered in such case.

Different other factors affect performance of an ultrasonic rangefinder. The size, composition, shape, and orientation of objects must be considered. In the cases presented in upper images in Fig. 7.16, measurements are normally correct, while in the scenarios presented in lower images in Fig. 7.16 the ultrasonic rangefinder would give false results.

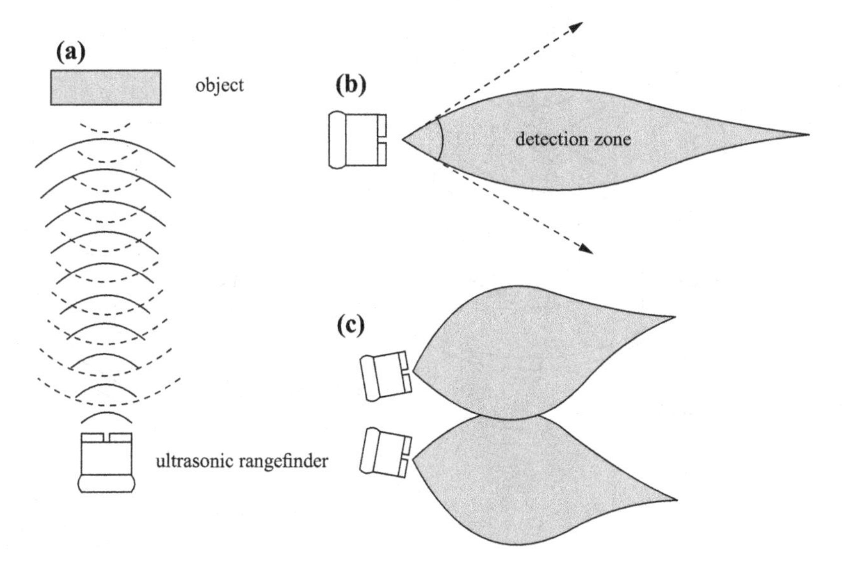

Fig. 7.15 Ultrasonic rangefinder: **a** principle of operation, **b** detection zone, and **c** combination of multiple sensors

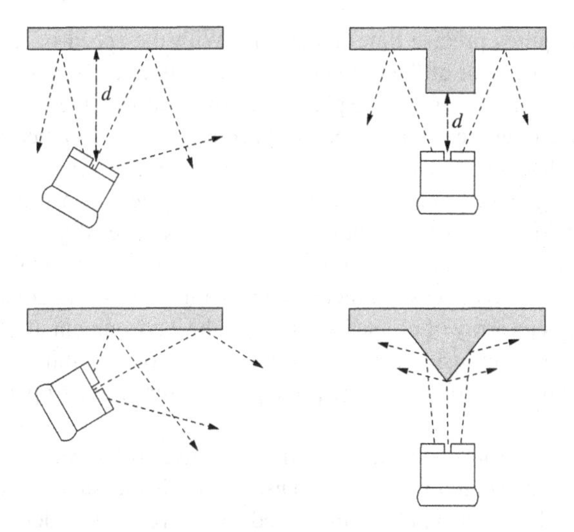

Fig. 7.16 Ultrasonic rangefinder distance measurement and limitations: correct measurements of distance d (upper row) and false results (lower row)

7.4.2 Laser Rangefinder and Laser Scanner

A laser rangefinder uses a laser beam to determine the distance to an object. The most common form of laser rangefinder operates on the time of flight principle. Distance can be determined by measuring the time it takes for the laser pulse to return to the

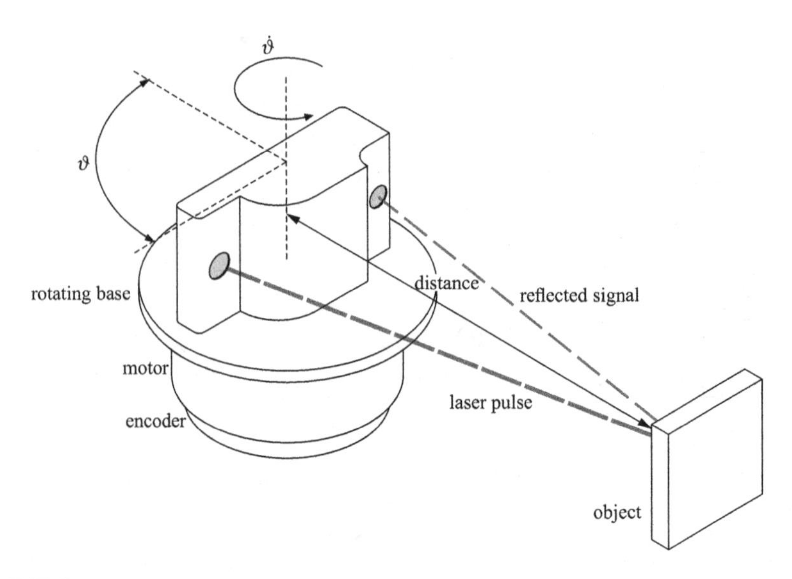

Fig. 7.17 Laser scanner

sensor and it requires precise time measuring. With the speed of light known and an accurate measurement of the time taken, the distance can be calculated. Another possibility is to compute light-wave phase shift by analyzing the incoming light and comparing it to a reference signal. The most accurate method to measure changes in distance rather than absolute distances is interferometry.

The laser rangefinder measures distance to one object at a time. Therefore, it is a one dimensional sensor. The laser scanner uses a laser that sweeps across the sensor's field of view. As the name implies, the instrument principally consists of a laser and a scanner. Distances are measured as with the laser rangefinder. The laser scanner produces an array of points by sampling the environment at a high rate. This is typically achieved by using rotating assemblies or rotating mirrors to sweep 360 degrees around the environment. The principle of laser scanner operation is shown in Fig. 7.17.

Sampled points represent object positions relative to the sensor. Generation of array of points is presented in Fig. 7.18. Distance d_L is measured by using the laser and rotation angle ϑ_L is typically measured by using an encoder on the rotating assembly. Points are therefore defined in polar coordinates. They can be transformed into Cartesian coordinates (x_L, y_L) relative to the sensor with

$$x_L = d_L \cos \varphi_L \quad \text{and} \quad y_L = d_L \sin \varphi_L. \tag{7.20}$$

Sampled points can be used to generate a map of the environment, for path planning, and avoidance of obstacles. A three-dimensional (3-D) scanner enables scanning of

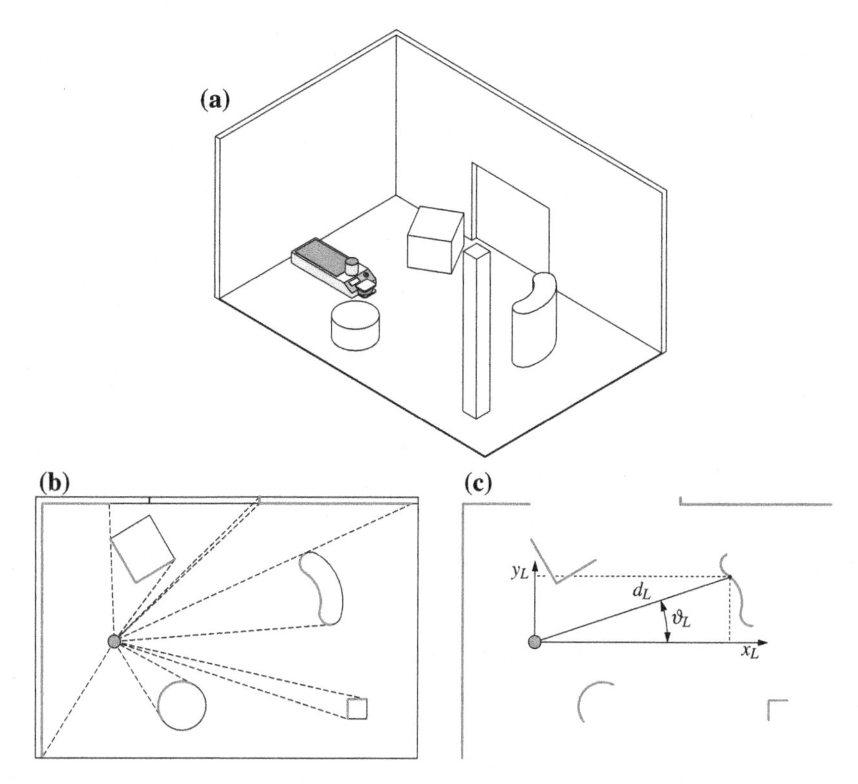

Fig. 7.18 Laser scanner used to create a map of environment: **a** environment, **b** scanning, and **c** map

a complete 3-D space and collecting a 3-D point cloud data by using another degree of freedom at the rotating assembly. These 3-D laser-scanners are typically termed LiDAR (Light Detection And Ranging) and are often used in autonomous vehicles for scanning the environment.

Chapter 8
Robot Vision

The task of robot vision is to recognize the geometry of the robot workspace from a digital image. It is our aim to find the relation between the coordinates of a point in the two-dimensional (2D) image and the coordinates of the point in the real three-dimensional (3D) robot environment.

8.1 System Configuration

The robot vision system is based on the use of one, two or more cameras. If several cameras are used to observe the same object, information about the depth of the object can be derived. In such case, we talk about 3D or stereo vision. Of course, the 3D view can also be achieved with a single camera if two images of the object are available, captured from different poses. If only one image is available, the depth can be estimated based on some previously known geometric properties of the object.

When analyzing the configuration of the robotic vision system, it is necessary to distinguish between possible placements of the cameras. The cameras can be placed in a fixed configuration, where they are rigidly mounted in the workcell, or in a mobile configuration, where the camera is attached to a robot. In the first configuration, the camera observes objects from a fixed position with respect to the robot base coordinate frame. The field of view of the camera does not change during the execution of the task, which means that basically the accuracy of the measurement is constant. In some tasks, it is difficult to prevent the manipulator from reaching into the field of view of the camera and thereby occluding the objects. Therefore, in such case, it is necessary to put a camera on a robot (in a mobile configuration).

The camera can be attached before or after the robot wrist. In the first case, the camera observes the situation from a favorable position and the manipulator generally does not occlude its field of view. In the second case, the camera is attached to the robot end-effector and typically only observes the object that is being manip-

M. Mihelj et al., *Robotics*, https://doi.org/10.1007/978-3-319-72911-4_8

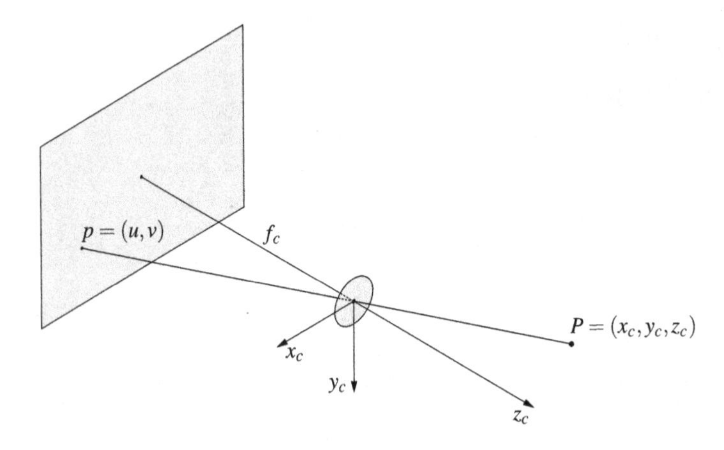

Fig. 8.1 Perspective projection

ulated. In both cases, the field of view of the camera changes with movements of the manipulator. When the manipulator approaches the object, the accuracy of the measurement typically increases.

8.2 Forward Projection

The basic equations of optics determine the position of a point in the image plane with respect to the corresponding point in 3D space (Fig. 8.1). We will therefore find the geometrical relation between the coordinates of the point $P = (x_c, y_c, z_c)$ in space and the coordinates of the point $p = (u, v)$ in the image.

As the aperture of the camera lenses, through which the light falls onto the image plane, is small compared to the size of the objects manipulated by the robot, we can replace the lenses in our mathematical model by a simple pinhole. In perspective projection points from space are projected onto the image plane by lines intersecting in a common point called the center of projection. When replacing a real camera with a pinhole camera, the center of projection is located in the center of the lenses.

When studying robot geometry and kinematics, we attached a coordinate frame to each rigid body (e.g., to robot segments or to objects manipulated by the robot). When considering robot vision, the camera itself represents a rigid body and a coordinate frame should be assigned to it. The pose of the camera will be from now on described by a corresponding coordinate frame. The z_c axis of the camera frame is directed along the optical axis, while the origin of the frame is positioned at the center of projection. We shall choose a right-handed frame where the x_c axis is parallel to the rows of the imaging sensor and the y_c axis is parallel with its columns.

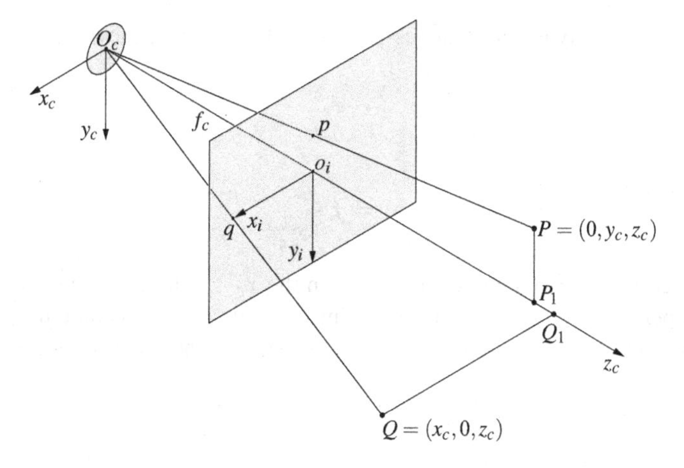

Fig. 8.2 Equivalent image plane

The image plane is in the camera, which is placed behind the center of projection. The distance f_c between the image and the center of projection is called the focal length. In the camera frame the focal length has a negative value, as the image plane intercepts the negative z_c axis. It is more convenient to use the equivalent image plane placed at a positive z_c value (Fig. 8.2). The equivalent image plane and the real image plane are symmetrical with respect to the origin of the camera frame. The geometrical properties of the objects are equivalent in both planes and differ only in the sign.

From now on we shall call the equivalent image plane simply the image plane. Also the image plane can be considered as a rigid body to which a coordinate frame should be attached. The origin of this frame is placed in the intersection of the optical axis with the image plane. The x_i and y_i axes are parallel to the x_c and y_c axes of the camera frame.

In this way the camera has two coordinate frames, the camera frame and the image frame. Let the point P be expressed in the camera frame, while the point p represents its projection onto the image plane. It is our aim to find the relations between the coordinates of the point P and the coordinates of its image p.

Let us first assume that the point P is located in the y_c–z_c plane of the camera frame. Its coordinates are

$$P = \begin{bmatrix} 0 \\ y_c \\ z_c \end{bmatrix}. \tag{8.1}$$

The projected point p is in this case located in the y_i axis of the image plane

$$p = \begin{bmatrix} 0 \\ y_i \end{bmatrix}. \tag{8.2}$$

Because of similarity of the triangles PP_1O_c and poO_c we can write

$$\frac{y_c}{y_i} = \frac{z_c}{f_c}$$

or

$$y_i = f_c \frac{y_c}{z_c}. \tag{8.3}$$

Let us consider also the point Q laying in the x_c–z_c plane of the camera frame. After the perspective projection of the point Q, its image q falls onto the x_i axis of the image frame. Because of similar triangles QQ_1O_c and qoO_c we have

$$\frac{x_c}{x_i} = \frac{z_c}{f_c}$$

or

$$x_i = f_c \frac{x_c}{z_c}. \tag{8.4}$$

In this way we obtained the relation between the coordinates (x_c, y_c, z_c), of the point P in the camera frame and the coordinates (x_i, y_i), of the point p in the image plane. Equations (8.3) and (8.4) represent the mathematical description of the perspective projection from a 3D onto a 2D space. Both equations can be written in the form of perspective matrix equation

$$s \begin{bmatrix} x_i \\ y_i \\ 1 \end{bmatrix} = \begin{bmatrix} f_c & 0 & 0 & 0 \\ 0 & f_c & 0 & 0 \\ 0 & 0 & 1 & 0 \end{bmatrix} \begin{bmatrix} x_c \\ y_c \\ z_c \\ 1 \end{bmatrix}. \tag{8.5}$$

In Eq. (8.5) s is a scaling factor, while (x_i, y_i) are the coordinates of the projected point in the image frame and (x_c, y_c, z_c) are the coordinates of the original point in the camera frame.

From the matrix Eq. (8.5) it is not difficult to realize that we can uniquely determine the coordinates (x_i, y_i) and the scaling factor s when knowing (x_c, y_c, z_c). On the contrary, we cannot calculate the coordinates (x_c, y_c, z_c) in the camera frame when only the coordinates (x_i, y_i) in the image frame are known, but not the scaling factor. Equation (8.5) represents the forward projection in robot vision. The calculation of (x_c, y_c, z_c) from (x_i, y_i) is called backward projection. When using a single camera and without a priori information about the size of the objects in the robot environment, a unique solution of the inverse problem cannot be found.

For ease of programming it is more convenient to use indices, marking the position of a pixel (i.e., the smallest element of a digital image) in a 2D image instead of metric units along the x_i and y_i axes of the image frame. We shall use two indices which we shall call index coordinates of a pixel (Fig. 8.3). These are the row index and the column index. In the memory storing the digital image the row index runs from the

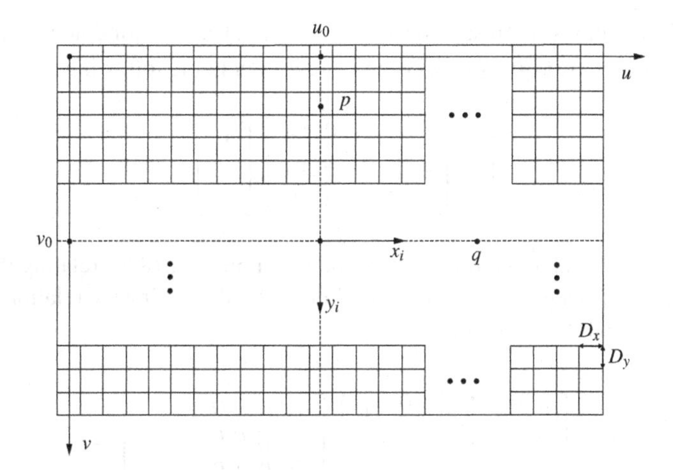

Fig. 8.3 The image plane and the index coordinate frame

top of the image to the bottom while the column index starts at the left and stops at the right edge of the image. We shall use the u axis for the column indices and the v axis for the row indices. In this way the index coordinate frame u–v belongs to each particular image. The upper left pixel is denoted either by $(0, 0)$, or $(1, 1)$. The index coordinates have no measuring units.

In the further text we shall find the relation between the image coordinates (x_i, y_i) and the index coordinates (u, v). Let us assume that the digital image was obtained as a direct output from the image sensor (A/D conversion was performed at the output of the image sensor). In this case each pixel corresponds to a particular element of the image sensor. We shall assume that the area of the image sensor is rectangular.

The origin of the image frame is in the point (u_0, v_0) of the index frame. The size of a pixel is represented by the pair (D_x, D_y). The relation between the image frame x_i–y_i and the index frame u–v is described by the following two equations

$$\begin{aligned} \frac{x_i}{D_x} &= u - u_0 \\ \frac{y_i}{D_y} &= v - v_0. \end{aligned} \tag{8.6}$$

Equations (8.6) can be rewritten as

$$\begin{aligned} u &= u_0 + \frac{x_i}{D_x} \\ v &= v_0 + \frac{y_i}{D_y}. \end{aligned} \tag{8.7}$$

In Eq. (8.7), $\frac{x_i}{D_x}$ and $\frac{y_i}{D_y}$ represent the number of digital conversions along the row and column respectively. Equation (8.7) can be rewritten in the following matrix form

$$\begin{bmatrix} u \\ v \\ 1 \end{bmatrix} = \begin{bmatrix} \frac{1}{D_x} & 0 & u_0 \\ 0 & \frac{1}{D_y} & v_0 \\ 0 & 0 & 1 \end{bmatrix} \begin{bmatrix} x_i \\ y_i \\ 1 \end{bmatrix}. \tag{8.8}$$

Using the pinhole camera model, we can now combine Eq. (8.5), relating the image coordinates to the camera coordinates, and Eq. (8.8), describing the relation between the image and index coordinates

$$\begin{aligned} s \begin{bmatrix} u \\ v \\ 1 \end{bmatrix} &= \begin{bmatrix} \frac{1}{D_x} & 0 & u_0 \\ 0 & \frac{1}{D_y} & v_0 \\ 0 & 0 & 1 \end{bmatrix} \begin{bmatrix} f_c & 0 & 0 & 0 \\ 0 & f_c & 0 & 0 \\ 0 & 0 & 1 & 0 \end{bmatrix} \begin{bmatrix} x_c \\ y_c \\ z_c \\ 1 \end{bmatrix} = \\ &= \begin{bmatrix} \frac{f_c}{D_x} & 0 & u_0 & 0 \\ 0 & \frac{f_c}{D_y} & v_0 & 0 \\ 0 & 0 & 1 & 0 \end{bmatrix} \begin{bmatrix} x_c \\ y_c \\ z_c \\ 1 \end{bmatrix}. \end{aligned} \tag{8.9}$$

The above matrix can be written also in the following form

$$\mathbf{P} = \begin{bmatrix} f_x & 0 & u_0 & 0 \\ 0 & f_y & v_0 & 0 \\ 0 & 0 & 1 & 0 \end{bmatrix}. \tag{8.10}$$

The **P** matrix represents the perspective projection from the camera frame into the corresponding index coordinate frame. The variables

$$\begin{aligned} f_x &= \frac{f_c}{D_x} \\ f_y &= \frac{f_c}{D_y} \end{aligned} \tag{8.11}$$

are the focal lengths of the camera along the x_c and y_c axes. The parameters f_x, f_y, u_0, and v_0 are called the intrinsic parameters of a camera.

In general the intrinsic parameters of the camera are not known. The specifications of the camera and the lenses are not sufficiently accurate. The intrinsic parameters of the camera are therefore obtained through the camera calibration process. When knowing the intrinsic parameters of the camera we can uniquely calculate the index coordinates (u, v) from the given coordinates (x_c, y_c, z_c). The coordinates (x_c, y_c, z_c) cannot be determined from the known (u, v) coordinates without knowing the scaling factor.

8.3 Backward Projection

The digital image is represented by a matrix of pixels. As the index coordinates (u, v) do not have measuring units, this means that characteristic features of the image are described more qualitatively than quantitatively. If we wish to express the distances in metric units, we must know the relation between the index coordinates (u, v) and the coordinates (x_r, y_r, z_r) in the 3D reference frame. Without knowing the real dimensions or the geometry of the scene it is impossible to recognize the features of the image.

8.3.1 *Single Camera*

Let us assume that we have a robot vision system with a single camera. The system has the image of the robot workspace as the input and is required to reproduce geometrical measurements as its output. The necessary transformations between the coordinate frames are evident from Fig. 8.4.

Let us suppose that we are now in a position to recognize the point q in the image. It is our aim to determine the coordinates of the real point Q from the coordinates of its image q. This is the problem of backward projection. In order to solve the problem, we must know how the coordinates of the point q are related to the coordinates of the real point Q in the reference frame, which is the problem of forward projection.

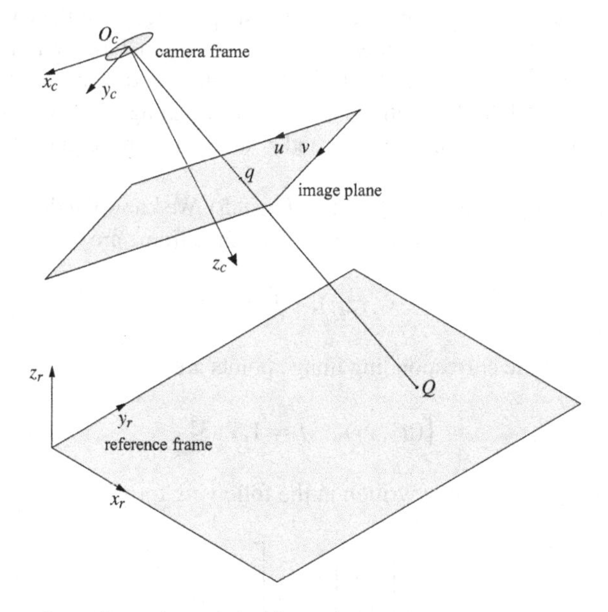

Fig. 8.4 The coordinate frames in a robot vision system

Let us solve first the problem of forward projection. The point Q is given by the coordinates (x_r, y_r, z_r) in the reference coordinate frame. We wish to determine the coordinates of its image $q = (u, v)$, expressed in the index frame. The frame x_c–y_c–z_c is attached to the camera. The matrix $\mathbf{M}$ represents the transformation from the reference into the camera frame

$$\begin{bmatrix} x_c \\ y_c \\ z_c \\ 1 \end{bmatrix} = \mathbf{M} \begin{bmatrix} x_r \\ y_r \\ z_r \\ 1 \end{bmatrix}. \tag{8.12}$$

By combining Eqs. (8.12) and (8.9), we obtain

$$s \begin{bmatrix} u \\ v \\ 1 \end{bmatrix} = \mathbf{PM} \begin{bmatrix} x_r \\ y_r \\ z_r \\ 1 \end{bmatrix}. \tag{8.13}$$

The relation (8.13) describes the forward projection. The elements of the $\mathbf{P}$ matrix are the intrinsic parameters of the camera, while the elements of the $\mathbf{M}$ matrix represent its extrinsic parameters. The 3×4 matrix

$$\mathbf{H} = \mathbf{PM} \tag{8.14}$$

is called the calibration matrix of the camera. It is used in the calibration process in order to determine both the intrinsic and extrinsic parameters of the camera.

In the further text we shall consider backward projection. It is our aim to determine the coordinates (x_r, y_r, z_r) of the real point Q from the known coordinates of the image point (u, v) and the calibration matrix $\mathbf{H}$. The scaling factor s is not known. In (8.13) we have four unknowns s, x_r, y_r, and z_r and only three equations for a single point in space.

Let us try with three points A, B, and C (Fig. 8.5). We know the distances between these three points. Their coordinates in the reference frame are

$$\left\{(x_{r_j}, y_{r_j}, z_{r_j}), \quad j = 1, 2, 3\right\}.$$

The coordinates of the corresponding image points are

$$\left\{(u_j, v_j), \quad j = 1, 2, 3\right\}.$$

The forward projection can be written in the following form

$$s_j \begin{bmatrix} u_j \\ v_j \\ 1 \end{bmatrix} = \mathbf{H} \begin{bmatrix} x_{r_j} \\ y_{r_j} \\ z_{r_j} \\ 1 \end{bmatrix}. \tag{8.15}$$

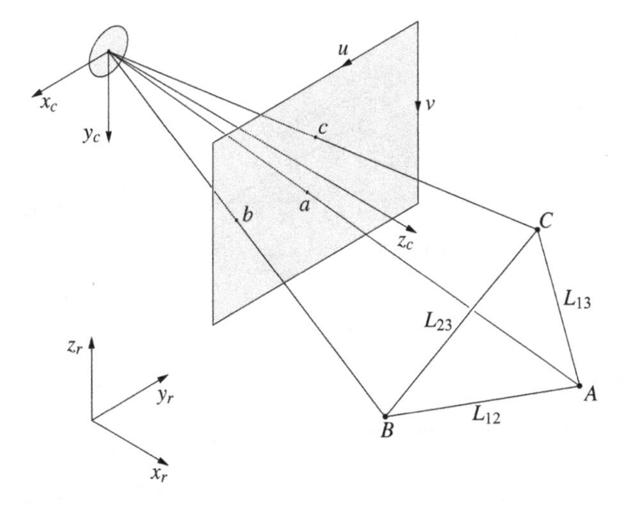

Fig. 8.5 Example of projecting three points

In Eq. (8.15) we have 12 unknowns and 9 equations. To solve the problem we need additional three equations. These equations can be obtained from the size of the triangle represented by the points A, B, and C. We shall denote the triangle sides AB, BC, and CA as the lengths L_{12}, L_{23}, and L_{31}

$$\begin{aligned} L_{12}^2 &= (x_{r_1} - x_{r_2})^2 + (y_{r_1} - y_{r_2})^2 + (z_{r_1} - z_{r_2})^2 \\ L_{23}^2 &= (x_{r_2} - x_{r_3})^2 + (y_{r_2} - y_{r_3})^2 + (z_{r_2} - z_{r_3})^2 \\ L_{31}^2 &= (x_{r_3} - x_{r_1})^2 + (y_{r_3} - y_{r_1})^2 + (z_{r_3} - z_{r_1})^2. \end{aligned} \tag{8.16}$$

Now we have twelve equations for the twelve unknowns. Thus, the solution of the inverse problem exists. It is inconvenient that the last three equations are nonlinear, requiring a computer for numerical solving of the equations. The approach is called model based backward projection.

8.3.2 *Stereo Vision*

Since the model of the observed object is usually not available or the object changes with time, other solutions to the backward projection problem need to be found. One possible solution is the use of stereo vision: sensing based on two cameras. The principle is similar to human visual perception where the images seen by the left and right eyes differ slightly due to parallax and the brain uses the differences between images to determine the distance to the observed object.

For simplicity we will assume two parallel cameras that observe point Q as shown in Fig. 8.6. Point Q is projected onto the image plane of the left and right cameras. The left camera's image plane contains projection q_l with coordinates $x_{i,l}$ and $y_{i,l}$

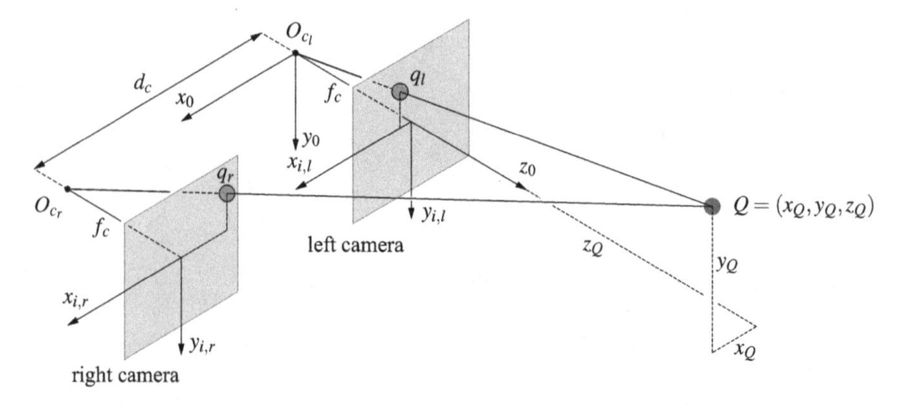

Fig. 8.6 Stereo view of point Q using two parallel cameras

while the right camera's image plane contains projection q_r with coordinates $x_{i,r}$ and $y_{i,r}$. The axes of the vision system coordinate frame x_0–y_0–z_0 have the same directions as the left camera's coordinate frame.

Figure 8.7a shows the top view, while Fig. 8.7b shows the side view of the situation in Fig. 8.6. These views will help us calculate the coordinates of point Q. From the geometry in Fig. 8.7a we can extract the following relations (distances x_Q, y_Q, and z_Q are with regard to the coordinate frame x_0–y_0–z_0)

$$\begin{aligned} \frac{z_Q}{f_c} &= \frac{x_Q}{x_{i,l}} \\ \frac{z_Q}{f_c} &= \frac{x_Q - d_c}{x_{i,r}}, \end{aligned} \tag{8.17}$$

where d_c is the distance between the cameras. From the first equation in (8.17) we express

$$x_Q = \frac{x_{i,l}}{f_c} z_Q \tag{8.18}$$

and insert into the second equation to get

$$\frac{x_{i,l} z_Q}{x_{i,r} f_c} - \frac{z_Q}{f_c} = \frac{d_c}{x_{i,r}}. \tag{8.19}$$

We can then determine the distance z_Q to point Q as

$$z_Q = \frac{f_c d_c}{x_{i,l} - x_{i,r}}. \tag{8.20}$$

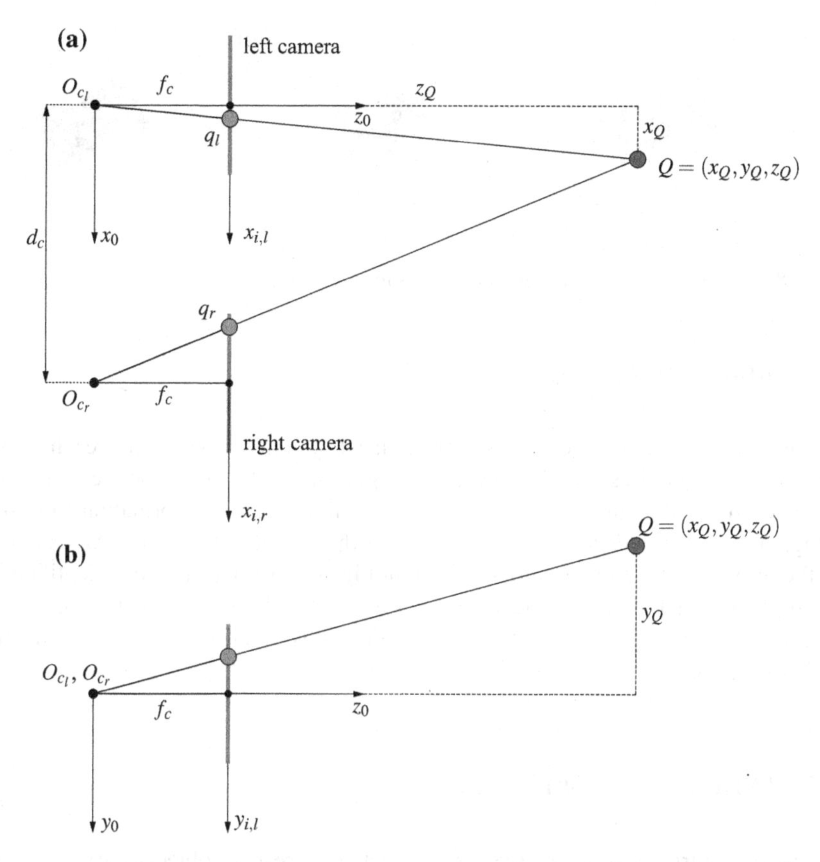

Fig. 8.7 Projections of point Q on the planes of the left and right cameras. The upper figure **a** shows a view of both cameras from above, while the lower figure **b** shows a side view of the cameras

The distance x_Q can be determined from Eq. (8.18). To determine distance y_Q we refer to Fig. 8.7b. From the geometry we can extract relation

$$\frac{z_Q}{f_c} = \frac{y_Q}{y_{i,l}}, \tag{8.21}$$

allowing us to calculate the remaining coordinate

$$y_Q = \frac{y_{i,l}}{f_c} z_Q. \tag{8.22}$$

Use of two cameras enables computation of the position (and orientation) of an object in space without an accurate model of the object.

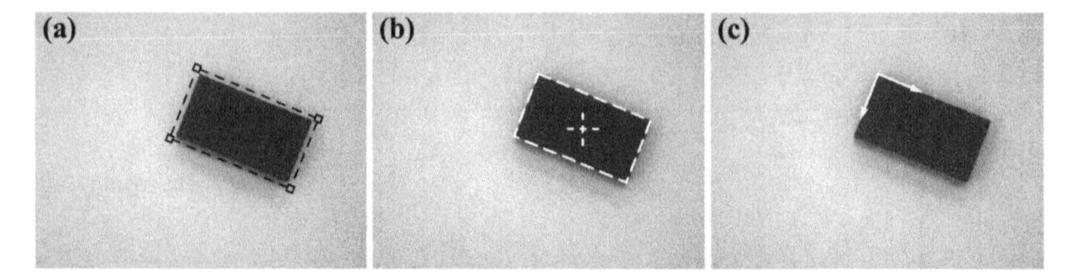

Fig. 8.8 **a** Model definition, **b** recognized object's features, **c** located object

8.4 Image Processing

In contrast to most other sensory systems, visual systems provide very reach information, which requires complex processing algorithms before it can be used for robot control. The goal of image processing is to obtain numerical information from the image, which provides a robust description of the object in the scene. An example of the result of image processing is shown in Fig. 8.8. An object is first identified in the scene and then its pose is determined as marked with the coordinate frame.

Image processing is beyond the scope of this book and it will not be specifically addressed here.

8.5 Object Pose from Image

In order to control the robot relative to the object of interest, the object pose needs to be defined relative to the robot coordinate frame x–y–z. As shown in Fig. 8.8, the pose of the object is known in the image coordinate frame after the image processing. In order to determine its pose in the robot frame, the transformation between the image and the robot coordinate frame must be defined, which is the result of camera calibration. Figure 8.9 presents a simple approach for the calibration problem, where the image plane is parallel to the horizontal plane. For simplicity, the image frame x_i–y_i–z_i is located at the same point as the index frame u–v (the z_i axis was added to the image frame to emphasize the rotation around the vertical axis).

8.5.1 Camera Calibration

Camera is mounted in a fixed position over the robot workspace. The calibration is performed with the calibration pattern (checkerboard), and the calibration tip at the robot end-effector. The calibration pattern can be augmented with a fiducial marker, which appears in the image for use as a point of reference or a measure. The goal of the calibration procedure is to find the transformation matrix $\mathbf{H}_i$ between the image

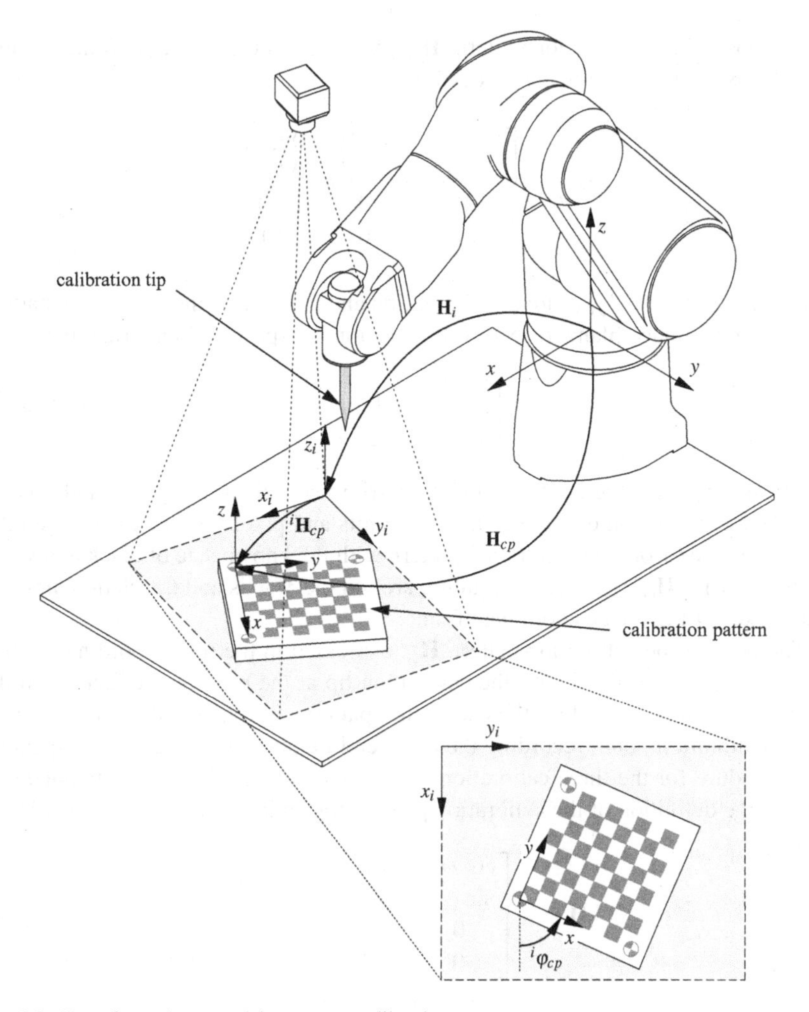

Fig. 8.9 Transformations used for camera calibration

and the robot coordinate frames x_i–y_i–z_i and x–y–z. Based on relations in Fig. 8.9 the following equality can be written

$$\mathbf{H}_{cp} = \mathbf{H}_i {}^i\mathbf{H}_{cp}, \tag{8.23}$$

where $\mathbf{H}_{cp}$ and ${}^i\mathbf{H}_{cp}$ are the poses of the calibration pattern expressed in the robot and the image coordinate frame, respectively.

The pose of the calibration pattern ${}^{i}\mathbf{H}_{cp}$ expressed in the image coordinate frame x_i–y_i–z_i is the result of image processing

$$
{}^{i}\mathbf{H}_{cp} = \begin{bmatrix} \cos{}^{i}\varphi_{cp} & -\sin{}^{i}\varphi_{cp} & 0 & {}^{i}x_{cp} \\ \sin{}^{i}\varphi_{cp} & \cos{}^{i}\varphi_{cp} & 0 & {}^{i}y_{cp} \\ 0 & 0 & 1 & 0 \\ 0 & 0 & 0 & 1 \end{bmatrix}, \tag{8.24}
$$

where ${}^{i}\varphi_{cp}$ and $({}^{i}x_{cp}, {}^{i}y_{cp})$ are the orientation and position of the calibration pattern relative to the image plane, respectively. Position is expressed in metric units as

$$
\begin{bmatrix} {}^{i}x_{cp} \\ {}^{i}y_{cp} \end{bmatrix} = \lambda \begin{bmatrix} u_{cp} \\ v_{cp} \end{bmatrix}, \tag{8.25}
$$

where (u_{cp}, v_{cp}) are the calibration pattern origin coordinates in pixels and λ is the ratio between position expressed in metric units and pixels on the image (the ratio can be obtained from the calibration pattern with the known size of black and white fields). Matrix ${}^{i}\mathbf{H}_{cp}$ represents a rotation around the z_i axis and translation along x_i and y_i axes of the image coordinate frame.

The pose of the calibration pattern $\mathbf{H}_{cp}$ expressed in the robot coordinate frame x–y–z can be determined with the calibration tip at the robot end-effector and the calibration points marked on the calibration pattern. By placing the calibration tip on the calibration point, recording the robot end-effector coordinates and repeating the procedure for the three calibration points, a set of coordinates is obtained that enables the definition of the calibration pattern pose relative to the robot coordinate frame as

$$
\mathbf{H}_{cp} = \begin{bmatrix} \cos\varphi_{cp} & -\sin\varphi_{cp} & 0 & x_{cp} \\ \sin\varphi_{cp} & \cos\varphi_{cp} & 0 & y_{cp} \\ 0 & 0 & 1 & z_{cp} \\ 0 & 0 & 0 & 1 \end{bmatrix}, \tag{8.26}
$$

where φ_{cp} and (x_{cp}, y_{cp}, z_{cp}) are the orientation and position of the calibration pattern relative to the robot frame, respectively.

From Eqs. (8.23), (8.24) and (8.26) the transformation matrix between the image and the robot coordinate frames can be obtained as

$$
\mathbf{H}_i = \mathbf{H}_{cp}{}^{i}\mathbf{H}_{cp}^{-1}. \tag{8.27}
$$

8.5.2 Object Pose

With the known $\mathbf{H}_i$, the object pose $\mathbf{H}_o$ relative to the robot coordinate frame can be determined as shown in Fig. 8.10.

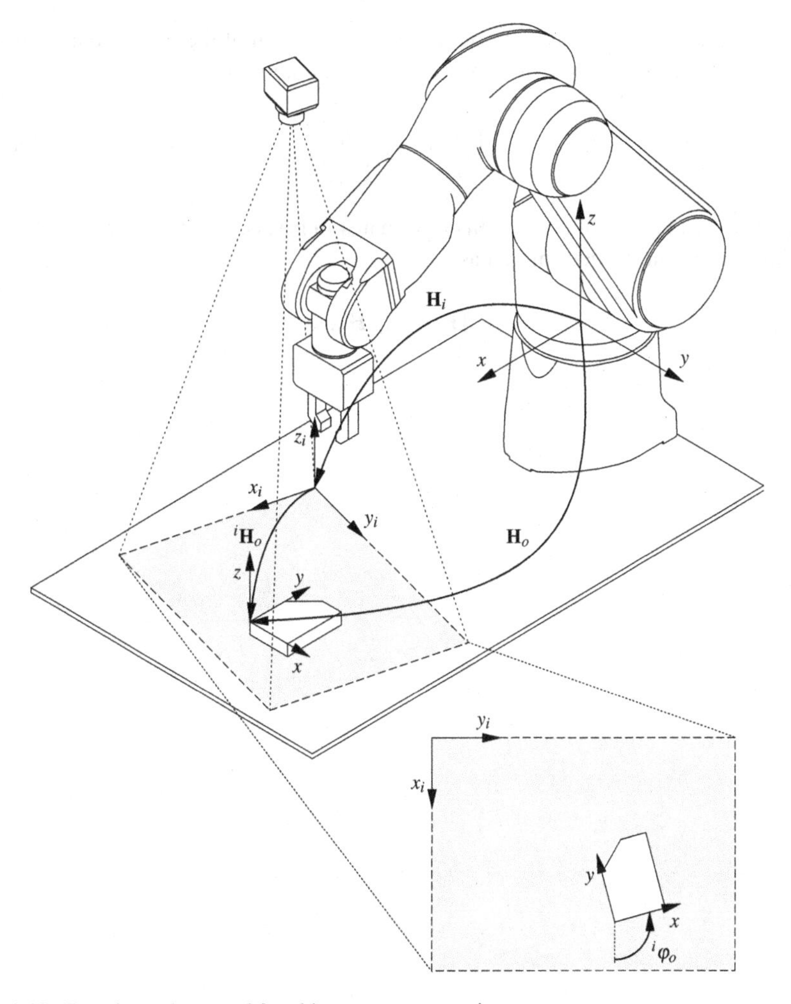

Fig. 8.10 Transformations used for object pose computation

The pose of the object ${}^i\mathbf{H}_o$ expressed in the image coordinate frame x_i–y_i–z_i is the result of image processing

$$
{}^i\mathbf{H}_o = \begin{bmatrix} \cos{}^i\varphi_o & -\sin{}^i\varphi_o & 0 & {}^i x_o \\ \sin{}^i\varphi_o & \cos{}^i\varphi_o & 0 & {}^i y_o \\ 0 & 0 & 1 & 0 \\ 0 & 0 & 0 & 1 \end{bmatrix}, \tag{8.28}
$$

where ${}^{i}\varphi_o$ and $({}^{i}x_o, {}^{i}y_o)$ are the orientation and position of the object relative to the image plane, respectively. Position is expressed in metric units as

$$\begin{bmatrix} {}^{i}x_o \\ {}^{i}y_o \end{bmatrix} = \lambda \begin{bmatrix} u_o \\ v_o \end{bmatrix}, \tag{8.29}$$

where (u_o, v_o) are the object origin coordinates in pixels.

Finally, $\mathbf{H}_o$ can be determined as

$$\mathbf{H}_o = \mathbf{H}_i {}^{i}\mathbf{H}_o. \tag{8.30}$$

Chapter 9
Trajectory Planning

In previous chapters we studied mathematical models of robot mechanisms. First of all we were interested in robot kinematics and dynamics. Before applying this knowledge to robot control, we must become familiar with the planning of robot motion. The aim of trajectory planning is to generate the reference inputs to the robot control system, which will ensure that the robot end-effector will follow the desired trajectory.

Robot motion is usually defined in the rectangular world coordinate frame placed in the robot workspace most convenient for the robot task. In the simplest task we only define the initial and the final point of the robot end-effector. The inverse kinematic model is then used to calculate the joint variables corresponding to the desired position of the robot end-effector.

9.1 Interpolation of the Trajectory Between Two Points

When moving between two points, the robot manipulator must be displaced from the initial to the final point in a given time interval t_f. Often we are not interested in the precise trajectory between the two points. Nevertheless, we must determine the time course of the motion for each joint variable and provide the calculated trajectory to the control input.

The joint variable is either the angle ϑ for the rotational or the displacement d for the translational joint. When considering the interpolation of the trajectory we shall not distinguish between the rotational and translational joints, so that the joint variable will be more generally denoted as q. With industrial manipulators moving between two points we most often select the so called trapezoidal velocity profile. The robot movement starts at $t = 0$ with constant acceleration, followed by the phase of constant velocity and finished by the constant deceleration phase (Fig. 9.1).

M. Mihelj et al., *Robotics*, https://doi.org/10.1007/978-3-319-72911-4_9

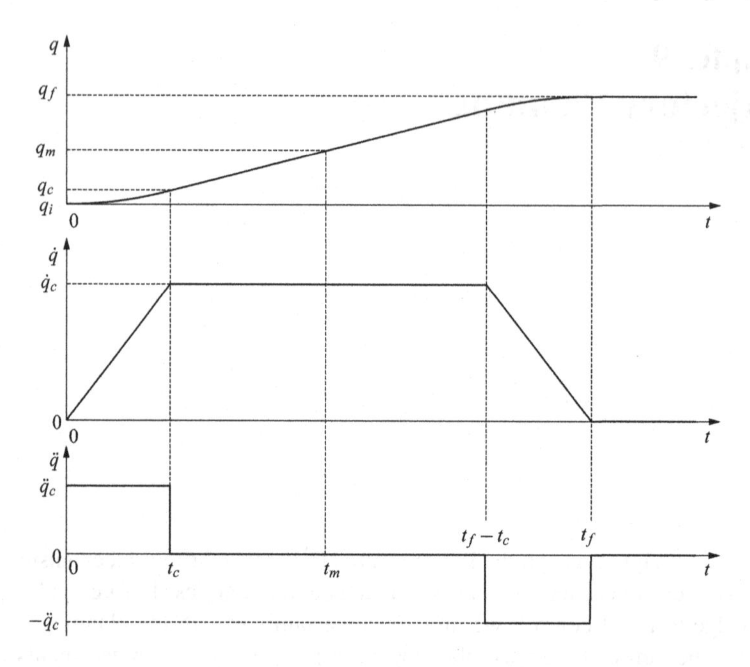

Fig. 9.1 The time dependence of the joint variables with trapezoidal velocity profile

The resulting trajectory of either the joint angle or displacement consists of the central linear interval, which is started and concluded with a parabolic segment. The initial and final velocities of the movement between the two points are zero. The duration of the constant acceleration phase is equal to the interval with the constant deceleration. In both phases the magnitude of the acceleration is $\ddot{q}_c$. In this way we deal with a symmetric trajectory, where

$$q_m = \frac{q_f + q_i}{2} \quad \text{at the moment} \quad t_m = \frac{t_f}{2}. \tag{9.1}$$

The trajectory $q(t)$ must satisfy several constraints in order that the robot joint will move from the initial point q_i to the final point q_f in the required time interval t_f. The velocity at the end of the initial parabolic phase must be equal to the constant velocity in the linear phase. The velocity in the first phase is obtained from the equation describing the constant acceleration motion

$$\dot{q} = \ddot{q}_c t. \tag{9.2}$$

At the end of the first phase we have

$$\dot{q}_c = \ddot{q}_c t_c. \tag{9.3}$$

The velocity in the second phase can be determined with the help of Fig. 9.1

$$\dot{q}_c = \frac{q_m - q_c}{t_m - t_c}, \tag{9.4}$$

where q_c represents the value of the joint variable q at the end of the initial parabolic phase (i.e., at the time t_c). Until that time the motion with constant acceleration $\ddot{q}_c$ takes place, so the velocity is determined by Eq. (9.2). The time dependence of the joint position is obtained by integrating Eq. (9.2)

$$q = \int \dot{q}dt = \ddot{q}_c \int tdt = \ddot{q}_c \frac{t^2}{2} + q_i, \tag{9.5}$$

where the initial joint position q_i is taken as the integration constant. At the end of the first phase we have

$$q_c = q_i + \frac{1}{2}\ddot{q}_c t_c^2. \tag{9.6}$$

The velocity at the end of the first phase (9.3) is equal to the constant velocity in the second phase (9.4)

$$\ddot{q}_c t_c = \frac{q_m - q_c}{t_m - t_c}. \tag{9.7}$$

By inserting Eq. (9.6) into Eq. (9.7) and considering the expression (9.1), we obtain, after rearrangement, the following quadratic equation

$$\ddot{q}_c t_c^2 - \ddot{q}_c t_f t_c + q_f - q_i = 0. \tag{9.8}$$

The acceleration $\ddot{q}_c$ is determined by the selected actuator and the dynamic properties of the robot mechanism. For chosen q_i, q_f, $\ddot{q}_c$, and t_f the time interval t_c is

$$t_c = \frac{t_f}{2} - \frac{1}{2}\sqrt{\frac{t_f^2 \ddot{q}_c - 4(q_f - q_i)}{\ddot{q}_c}}. \tag{9.9}$$

To generate the movement between the initial q_i and the final position q_f the following polynomial must be generated in the first phase

$$q(t) = q_i + \frac{1}{2}\ddot{q}_c t^2 \qquad 0 \le t \le t_c. \tag{9.10}$$

In the second phase a linear trajectory must be generated starting in the point (t_c, q_c), with the slope $\dot{q}_c$

$$(q - q_c) = \dot{q}_c(t - t_c). \tag{9.11}$$

After rearrangement we obtain

$$q(t) = q_i + \ddot{q}_c t_c (t - \frac{t_c}{2}) \qquad t_c < t \leq (t_f - t_c). \tag{9.12}$$

In the last phase the parabolic trajectory must be generated similarly to the first phase, only that now the extreme point is in (t_f, q_f) and the curve is turned upside down

$$q(t) = q_f - \frac{1}{2}\ddot{q}_c (t - t_f)^2 \qquad (t_f - t_c) < t \leq t_f. \tag{9.13}$$

In this way we obtained analytically the time dependence of the angle or displacement of the rotational or translational joint moving from point to point.

9.2 Interpolation by Use of via Points

In some robot tasks, movements of the end-effector more complex than point to point motions, are necessary. In welding, for example, the curved surfaces of the objects must be followed. Such trajectories can be obtained by defining, besides the initial and the final point, also the so called via points through which the robot end-effector must move.

In this chapter we shall analyze the problem, where we wish to interpolate the trajectory through n via points $\{q_1, \ldots, q_n\}$, which must be reached by the robot in time intervals $\{t_1, \ldots, t_n\}$. The interpolation will be performed with the help of trapezoidal velocity profiles. The trajectory will consist of a sequence of linear segments describing the movements between two via points and parabolic segments representing the transitions through the via points. In order to avoid the discontinuity of the first derivative at the moment t_k, the trajectory $q(t)$ must have a parabolic course in the vicinity of q_k. By doing so the second derivative in the point q_k (acceleration) remains discontinuous.

The interpolated trajectory, defined as a sequence of linear functions with parabolic transitions through the via points (the transition time Δt_k), is analytically described by the following constraints

$$q(t) = \begin{cases} a_{1,k}(t - t_k) + a_{0,k} & t_k + \frac{\Delta t_k}{2} \leq t < t_{k+1} - \frac{\Delta t_{k+1}}{2} \\ b_{2,k}(t - t_k)^2 + b_{1,k}(t - t_k) + b_{0,k} & t_k - \frac{\Delta t_k}{2} \leq t < t_k + \frac{\Delta t_k}{2}. \end{cases} \tag{9.14}$$

The coefficients $a_{0,k}$ and $a_{1,k}$ determine the linear parts of the trajectory, where k represents the index of the corresponding linear segment. The coefficients $b_{0,k}$, $b_{1,k}$ and $b_{2,k}$ belong to the parabolic transitions. The index k represents the consecutive number of a parabolic segment.

First, the velocities in the linear segments will be calculated from the given positions and the corresponding time intervals, as shown in Fig. 9.2. We assume that the initial and final velocities are equal to zero. In this case we have

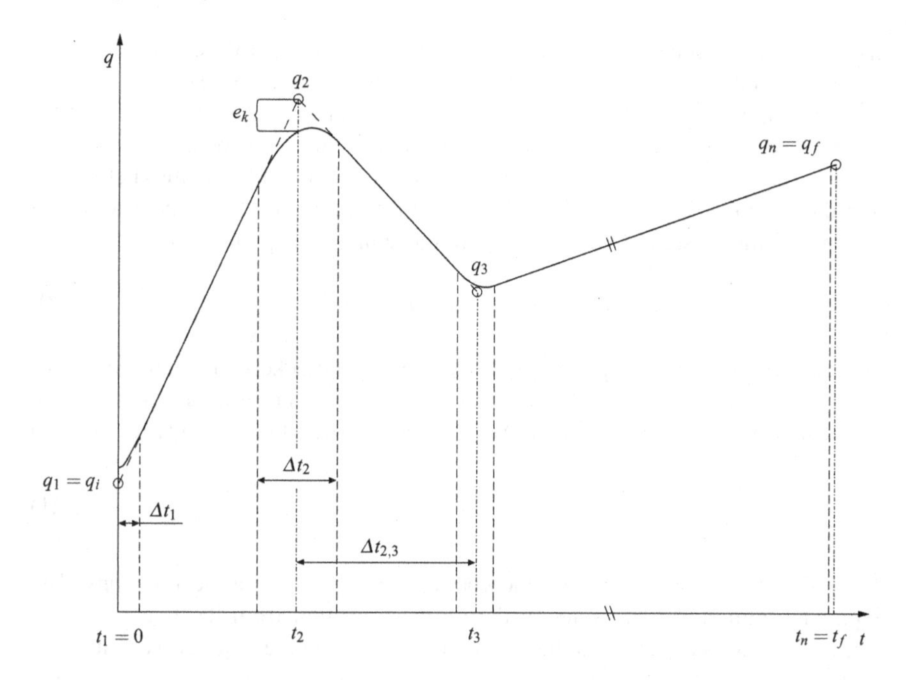

Fig. 9.2 Trajectory interpolation through n via points—linear segments with parabolic transitions are used

$$\dot{q}_{k-1,k} = \begin{cases} 0 & k = 1 \\ \frac{q_k - q_{k-1}}{t_k - t_{k-1}} & k = 2, \ldots, n \\ 0 & k = n + 1. \end{cases} \tag{9.15}$$

Further, we must determine the coefficients of the linear segments $a_{0,k}$ and $a_{1,k}$. The coefficient $a_{0,k}$ can be found from the linear function (9.14), by taking into account the known position at the moment t_k, when the robot segment approaches the point q_k

$$q(t_k) = q_k = a_{1,k}(t_k - t_k) + a_{0,k} = a_{0,k}, \tag{9.16}$$

therefore

$$t = t_k \quad \Rightarrow \quad a_{0,k} = q_k \qquad k = 1, \ldots, n-1\,. \tag{9.17}$$

The coefficient $a_{1,k}$ can be determined from the time derivative of the linear function (9.14)

$$\dot{q}(t) = a_{1,k}. \tag{9.18}$$

By considering the given velocities in the time interval $t_{k,k+1}$, we obtain

$$a_{1,k} = \dot{q}_{k,k+1} \qquad k = 1, \ldots, n-1. \tag{9.19}$$

In this way the coefficients of the linear segments of the trajectory are determined and we can proceed with the coefficients of the parabolic functions. We shall assume that the transition time is predetermined as Δt_k. If the transition time is not prescribed, the absolute value of the acceleration $|\ddot{q}_k|$ in the via point must be first defined and then the transition time is calculated from the accelerations and velocities before and after the via point. In this case only the sign of the acceleration must be determined by considering the sign of the velocity difference in the via point

$$\ddot{q}_k = sign(\dot{q}_{k,k+1} - \dot{q}_{k-1,k})|\ddot{q}_k|, \tag{9.20}$$

where $sign(\cdot)$ means the sign of the expression in the brackets. Given the values of the accelerations in the via points and the velocities before and after the via point, the time of motion through the via point Δt_k is calculated (deceleration and acceleration)

$$\Delta t_k = \frac{\dot{q}_{k,k+1} - \dot{q}_{k-1,k}}{\ddot{q}_k}. \tag{9.21}$$

We shall proceed by calculating the coefficients of the quadratic functions. The required continuity of the velocity during the transition from the linear into the parabolic trajectory segment at the instant $(t_k - \frac{\Delta t_k}{2})$ and the required velocity continuity during the transition from the parabolic into the linear segment at $(t_k + \frac{\Delta t_k}{2})$ represents the starting point for the calculation of the coefficients $b_{1,k}$ and $b_{2,k}$. First, we calculate the time derivative of the quadratic function (9.14)

$$\dot{q}(t) = 2b_{2,k}(t - t_k) + b_{1,k}. \tag{9.22}$$

Assuming that the velocity at the instant $(t_k - \frac{\Delta t_k}{2})$, is $\dot{q}_{k-1,k}$, while at $(t_k + \frac{\Delta t_k}{2})$, it is $\dot{q}_{k,k+1}$, we can write

$$\begin{aligned}
\dot{q}_{k-1,k} &= 2b_{2,k}\left(t_k - \frac{\Delta t_k}{2} - t_k\right) + b_{1,k} = -b_{2,k}\Delta t_k + b_{1,k} \qquad t = t_k - \frac{\Delta t_k}{2} \\
\dot{q}_{k,k+1} &= 2b_{2,k}\left(t_k + \frac{\Delta t_k}{2} - t_k\right) + b_{1,k} = b_{2,k}\Delta t_k + b_{1,k} \qquad t = t_k + \frac{\Delta t_k}{2}.
\end{aligned} \tag{9.23}$$

By adding Eq. (9.23), the coefficient $b_{1,k}$ can be determined

$$b_{1,k} = \frac{\dot{q}_{k,k+1} + \dot{q}_{k-1,k}}{2} \qquad k = 1, \ldots, n \tag{9.24}$$

and by subtracting Eq. (9.23), the coefficient $b_{2,k}$ is calculated

$$b_{2,k} = \frac{\dot{q}_{k,k+1} - \dot{q}_{k-1,k}}{2\Delta t_k} = \frac{\ddot{q}_k}{2} \qquad k = 1, \ldots, n. \tag{9.25}$$

By taking into account the continuity of the position at the instant $(t_k + \frac{\Delta t_k}{2})$, the coefficient $b_{0,k}$ of the quadratic polynomial can be calculated. At $(t_k + \frac{\Delta t_k}{2})$, the position $q(t)$, calculated from the linear function

$$q\left(t_k + \frac{\Delta t_k}{2}\right) = a_{1,k}\left(t_k + \frac{\Delta t_k}{2} - t_k\right) + a_{0,k} = \dot{q}_{k,k+1}\frac{\Delta t_k}{2} + q_k \tag{9.26}$$

equals the position $q(t)$, calculated from the quadratic function

$$\begin{aligned} q\left(t_k + \frac{\Delta t_k}{2}\right) &= b_{2,k}\left(t_k + \frac{\Delta t_k}{2} - t_k\right)^2 + b_{1,k}\left(t_k + \frac{\Delta t_k}{2} - t_k\right) + b_{0,k} \\ &= \frac{\dot{q}_{k,k+1} - \dot{q}_{k-1,k}}{2\Delta t_k}\left(\frac{\Delta t_k}{2}\right)^2 + \frac{\dot{q}_{k,k+1} + \dot{q}_{k-1,k}}{2}\cdot\frac{\Delta t_k}{2} + b_{0,k}\,. \end{aligned} \tag{9.27}$$

By equating (9.26) and (9.27) the coefficient $b_{0,k}$ is determined

$$b_{0,k} = q_k + (\dot{q}_{k,k+1} - \dot{q}_{k-1,k})\frac{\Delta t_k}{8}\,. \tag{9.28}$$

It can be verified that the calculated coefficient $b_{0,k}$ ensures also continuity of position at the instant $(t_k - \frac{\Delta t_k}{2})$. Such a choice of the coefficient $b_{0,k}$ prevents the joint trajectory going through point q_k. The robot only more or less approaches this point. The distance of the calculated trajectory from the reference point depends mainly on the decelerating and accelerating time interval Δt_k, which is predetermined by the required acceleration $|\ddot{q}_k|$. The error e_k of the calculated trajectory can be estimated by comparing the desired position q_k with the actual position $q(t)$ at the instant t_k, which is obtained by inserting t_k into the quadratic function (9.14)

$$e_k = q_k - q(t_k) = q_k - b_{0,k} = -(\dot{q}_{k,k+1} - \dot{q}_{k-1,k})\frac{\Delta t_k}{8}\,. \tag{9.29}$$

It can be noticed that the error e_k equals zero only when the velocities of the linear segments before and after the via points are equal or when the time interval Δt_k is zero, meaning infinite acceleration (which in reality is not possible).

The described approach to the trajectory interpolation has a minor deficiency. From Eq. (9.29) it can be observed that, instead of reaching the via point, the robot goes around it. As the initial and final trajectory points are also considered as via points, an error is introduced into the trajectory planning. At the starting point of the trajectory, the actual and the desired position differ by the error e_1 (Fig. 9.3, the light curve shows the trajectory without correction), arising from Eq. (9.29). The error represents a step in the position signal, which is not desired in robotics. To avoid this abrupt change in position, the first and the last trajectory point must be handled separately from the via points.

The required velocities in the starting and the final points should be zero. The velocity at the end of the time interval Δt_1 must be equal to the velocity in the first

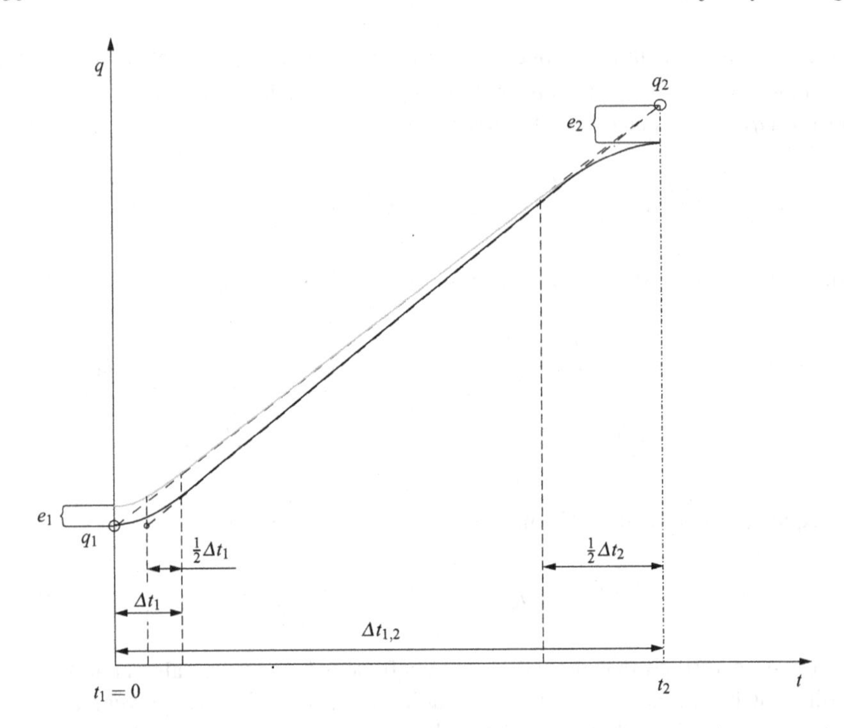

Fig. 9.3 Trajectory interpolation—enlarged presentation of the first segment of the trajectory shown in Fig. 9.2. The lighter curve represents the trajectory without correction, while the darker curve shows the corrected trajectory

linear segment. First, we calculate the velocity in the linear part

$$\dot{q}_{1,2} = \frac{q_2 - q_1}{t_2 - t_1 - \frac{1}{2}\Delta t_1}. \tag{9.30}$$

Equation (9.30) is similar to Eq. (9.15), only that now $\frac{1}{2}\Delta t_1$ is subtracted in the denominator, as in the short time interval (the beginning of the parabolic segment in Fig. 9.3) the position of the robot changes only to a very small extent. By doing so, a higher velocity in the linear segment of the trajectory is obtained. At the end of the acceleration interval Δt_1 we have

$$\frac{q_2 - q_1}{t_2 - t_1 - \frac{1}{2}\Delta t_1} = \ddot{q}_1 \Delta t_1 \tag{9.31}$$

We must determine also the acceleration $\ddot{q}_1$ at the starting point of the trajectory. Assuming that its absolute value $|\ddot{q}_1|$ was predetermined, only the sign must be adequately selected. The choice of the sign will be performed on the basis of the positional difference. In principle the velocity difference should be taken into account when determining the sign of acceleration, however the initial velocity is zero, and the sign can therefore depend on the difference in positions.

$$\ddot{q}_1 = sign(q_2 - q_1)|\ddot{q}_1|. \tag{9.32}$$

From Eq. (9.31), the time interval Δt_1 is calculated

$$(q_2 - q_1) = \ddot{q}_1 \Delta t_1 (t_2 - t_1 - \frac{1}{2}\Delta t_1). \tag{9.33}$$

After rearrangement we obtain

$$-\frac{1}{2}\ddot{q}_1 \Delta t_1^2 + \ddot{q}_1 (t_2 - t_1)\Delta t_1 - (q_2 - q_1) = 0, \tag{9.34}$$

so the time interval Δt_1 is

$$\Delta t_1 = \frac{-\ddot{q}_1 (t_2 - t_1) \pm \sqrt{\ddot{q}_1^2 (t_2 - t_1)^2 - 2\ddot{q}_1 (q_2 - q_1)}}{-\ddot{q}_1}, \tag{9.35}$$

and after simplifying Eq. (9.35)

$$\Delta t_1 = (t_2 - t_1) - \sqrt{(t_2 - t_1)^2 - \frac{2(q_2 - q_1)}{\ddot{q}_1}}. \tag{9.36}$$

In Eq. (9.36), the minus sign was selected before the square root, because the time interval Δt_1 must be shorter than $(t_2 - t_1)$. From Eq. (9.30), the velocity in the linear part of the trajectory can be calculated. As is evident from Fig. 9.3 (the darker curve represents the corrected trajectory), the introduced correction eliminates the error in the initial position.

Similarly, as for the first segment, the correction must be calculated also for the last segment between points q_{n-1} and q_n. The velocity in the last linear segment is

$$\dot{q}_{n-1,n} = \frac{q_n - q_{n-1}}{t_n - t_{n-1} - \frac{1}{2}\Delta t_n}. \tag{9.37}$$

In the denominator of Eq. (9.37) the value $\frac{1}{2}\Delta t_n$ was subtracted, as immediately before the complete stop of the robot, its position changes only very little. At the transition from the last linear segment into the last parabolic segment the velocities are equal

$$\frac{q_n - q_{n-1}}{t_n - t_{n-1} - \frac{1}{2}\Delta t_n} = \ddot{q}_n \Delta t_n. \tag{9.38}$$

The acceleration (deceleration) of the last parabolic segment is determined on the basis of the positional difference

$$\ddot{q}_n = sign(q_{n-1} - q_n)|\ddot{q}_n|. \tag{9.39}$$

By inserting the above equation into Eq. (9.38), we calculate, in a similar way as for the first parabolic segment, also the duration of the last parabolic segment

$$\Delta t_n = (t_n - t_{n-1}) - \sqrt{(t_n - t_{n-1})^2 - \frac{2(q_n - q_{n-1})}{\ddot{q}_n}}. \tag{9.40}$$

From Eq. (9.37), the velocity of the last linear segment can be determined. By considering the corrections at the start and at the end of the trajectory, the time course through the via points is calculated. In this way the entire trajectory was interpolated at the n points.

Chapter 10
Robot Control

The problem of robot control can be explained as a computation of the forces or torques which must be generated by the actuators in order to successfully accomplish the robot's task. The appropriate working conditions must be ensured both during the transient period as well as in the stationary state. The robot task can be presented either as the execution of the motions in a free space, where position control is performed, or in contact with the environment, where control of the contact force is required. First, we shall study the position control of a robot mechanism which is not in contact with its environment. Then, in the further text we shall upgrade the position control with the force control.

The problem of robot control is not unique. There exist various methods which differ in their complexity and in the effectiveness of robot actions. The choice of the control method depends on the robot task. An important difference is, for example, between the task where the robot end-effector must accurately follow the prescribed trajectory (e.g., laser welding) and another task where it is only required that the robot end-effector reaches the desired final pose, while the details of the trajectory between the initial and the final point are not important (e.g., palletizing). The mechanical structure of the robot mechanism also influences the selection of the appropriate control method. The control of a cartesian robot manipulator in general differs from the control of an anthropomorphic robot.

Robot control usually takes place in the world coordinate frame, which is defined by the user and is called also the coordinate frame of the robot task. Instead of world coordinate frame we often use a shorter expression, namely external coordinates. We are predominantly interested in the pose of the robot end-effector expressed in the external coordinates and rarely in the joint positions, which are also called internal coordinates. Nevertheless, we must be aware that in all cases we directly control the internal coordinates (i.e., joint angles or displacements). The end-effector pose is only controlled indirectly. It is determined by the kinematic model of the robot mechanism and the given values of the internal coordinates.

M. Mihelj et al., *Robotics*, https://doi.org/10.1007/978-3-319-72911-4_10

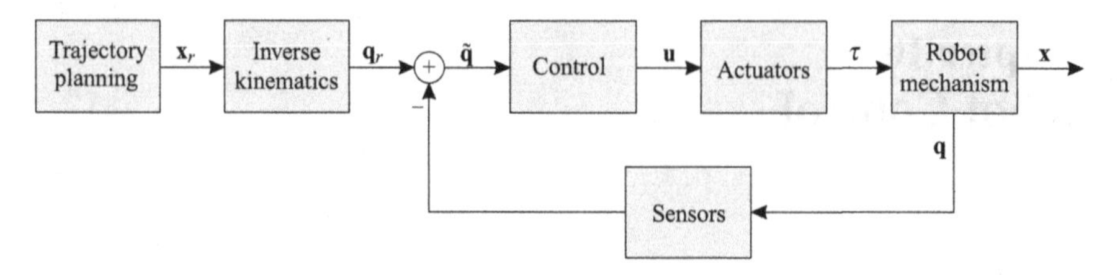

Fig. 10.1 A general robot control system

Figure 10.1 shows a general robot control system. The input to the control system is the desired pose of the robot end-effector, which is obtained by using trajectory interpolation methods, introduced in the previous chapter. The variable $\mathbf{x}_r$ represents the desired (i.e., the reference pose) of the robot end-effector. The $\mathbf{x}$ vector, describing the actual pose of the robot end-effector, in general comprises six variables. Three of them define the position of the robot end-point, while the other three determine the orientation of the robot end-effector. Thus, we write $\mathbf{x} = \left[x\ y\ z\ \varphi\ \vartheta\ \psi\right]^T$.

The position of the robot end-effector is determined by the vector from the origin of the world coordinate frame to the robot end-point. The orientation of the end-effector can be presented in various ways. One of the possible descriptions is the so called RPY notation, arising from aeronautics and shown in Fig. 4.4. The orientation is determined by the angle φ around the z axis (Roll), the angle ϑ around the y axis (Pitch), and the angle ψ around the x axis (Yaw).

By the use of the inverse kinematics algorithm the internal coordinates $\mathbf{q}_r$, corresponding to the desired end-effector pose, are calculated. The variable q_r represents the joint position (i.e., the angle ϑ for the rotational joint and the distance d for the translational joint). The desired internal coordinates are compared to the actual internal coordinates in the robot control system. On the basis of the positional error $\tilde{\mathbf{q}} = \mathbf{q}_r - \mathbf{q}$ the control system output $\mathbf{u}$ is calculated. The output $\mathbf{u}$ is converted from a digital into an analogue signal, amplified and delivered to the robot actuators. The actuators ensure the forces or torques necessary for the required robot motion. The robot motion is assessed by the sensors which were described in the chapter devoted to robot sensors.

10.1 Control of the Robot in Internal Coordinates

The simplest robot control approach is based on controllers where the control loop is closed separately for each particular degree of freedom. Such controllers are suitable for control of independent second order systems with constant inertial and damping parameters. This approach is less suitable for robotic systems characterized by nonlinear and time varying behavior.

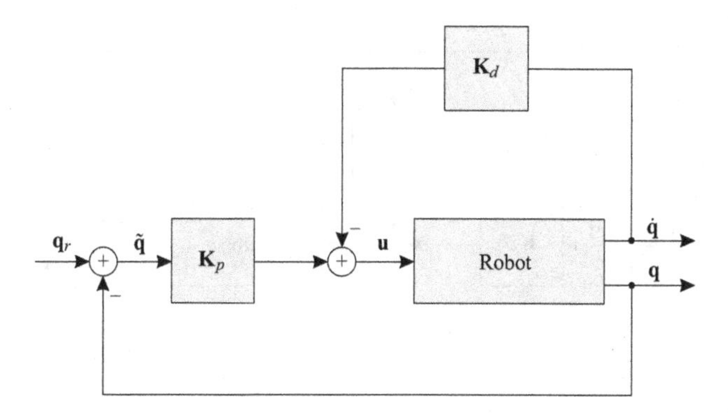

Fig. 10.2 PD position control with high damping

10.1.1 PD Control of Position

First, a simple proportional-derivative (PD) controller will be analyzed. The basic control scheme is shown in Fig. 10.2. The control is based on calculation of the positional error and determination of control parameters, which enable reduction or suppression of the error. The positional error is reduced for each joint separately, which means that as many controllers are to be developed as there are degrees of freedom. The reference positions $\mathbf{q}_r$ are compared to the actual positions of the robot joints $\mathbf{q}$

$$\tilde{\mathbf{q}} = \mathbf{q}_r - \mathbf{q}. \tag{10.1}$$

The positional error $\tilde{\mathbf{q}}$ is amplified by the proportional position gain $\mathbf{K}_p$. As a robot manipulator has several degrees of freedom, the error $\tilde{\mathbf{q}}$ is expressed as a vector, while $\mathbf{K}_p$ is a diagonal matrix of the gains of all joint controllers. The calculated control input provokes robot motion in the direction of reduction of the positional error. As the actuation of the robot motors is proportional to the error, it can happen that the robot will overshoot instead of stopping in the desired position. Such overshoots are not allowed in robotics, as they may result in collisions with objects in the robot vicinity. To ensure safe and stable robot actions, a velocity closed loop is introduced with a negative sign. The velocity closed loop brings damping into the system. It is represented by the actual joint velocities $\dot{\mathbf{q}}$ multiplied by a diagonal matrix of velocity gains $\mathbf{K}_d$. The control law can be written in the following form

$$\mathbf{u} = \mathbf{K}_p(\mathbf{q}_r - \mathbf{q}) - \mathbf{K}_d\dot{\mathbf{q}}, \tag{10.2}$$

where $\mathbf{u}$ represents the control inputs (i.e., the joint forces or torques), which must be provided by the actuators. From Eq. (10.2) we can notice that at higher velocities of robot motions, the velocity control loop reduces the joint actuation and, by damping the system, ensures robot stability.

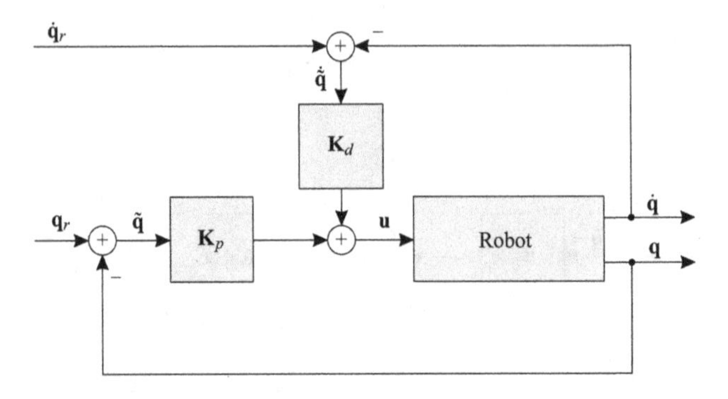

Fig. 10.3 PD position control

The control method shown in Fig. 10.2 provides high damping of the system in the fastest part of the trajectory, which is usually not necessary. Such behavior of the controller can be avoided by upgrading the PD controller with the reference velocity signal. This signal is obtained as the numerical derivative of the desired position. The velocity error is used as control input

$$\dot{\tilde{\mathbf{q}}} = \dot{\mathbf{q}}_r - \dot{\mathbf{q}}. \tag{10.3}$$

The control algorithm demonstrated in Fig. 10.3 can be written as

$$\mathbf{u} = \mathbf{K}_p(\mathbf{q}_r - \mathbf{q}) + \mathbf{K}_d(\dot{\mathbf{q}}_r - \dot{\mathbf{q}}). \tag{10.4}$$

As the difference between the reference velocity $\dot{\mathbf{q}}_r$ and $\dot{\mathbf{q}}$ is used instead of the total velocity $\dot{\mathbf{q}}$, the damping effect is reduced. For a positive difference the control loop can even accelerate the robot motion.

The synthesis of the PD position controller consists of determining the matrices $\mathbf{K}_p$ and $\mathbf{K}_d$. For fast response, the $\mathbf{K}_p$ gains must be high. By proper choice of the $\mathbf{K}_d$ gains, critical damping of the robot systems is obtained. The critical damping ensures fast response without overshoot. Such controllers must be built for each joint separately. The behavior of each controller is entirely independent of the controllers belonging to the other joints of the robot mechanism.

10.1.2 PD Control of Position with Gravity Compensation

In the chapter on robot dynamics we found that the robot mechanism is under the influence of inertial, Coriolis, centripetal, and gravitational forces (5.56). In general, friction forces occurring in robot joints, must also be included in the robot dynamic model. In a somewhat simplified model, only viscous friction, being proportional to the joint velocity, will be taken into account ($\mathbf{F}_v$ is a diagonal matrix of the joint friction coefficients). The enumerated forces must be overcome by the robot actuators, which is evident from the following equation, similar to Eq. (5.56)

$$\mathbf{B}(\mathbf{q})\ddot{\mathbf{q}} + \mathbf{C}(\mathbf{q}, \dot{\mathbf{q}})\dot{\mathbf{q}} + \mathbf{F}_v\dot{\mathbf{q}} + \mathbf{g}(\mathbf{q}) = \tau. \tag{10.5}$$

When developing the PD controller, we did not pay attention to the specific forces influencing the robot mechanism. The robot controller calculated the required actuation forces solely on the basis of the difference between the desired and the actual joint positions. Such a controller cannot predict the force necessary to produce the desired robot motion. As the force is calculated from the positional error, this means that in general the error is never equal to zero. When knowing the dynamic robot model, we can predict the forces which are necessary for the performance of a particular robot motion. These forces are then generated by the robot motors regardless of the positional error signal.

In quasi-static conditions, when the robot is standing still or moving slowly, we can assume zero accelerations $\ddot{\mathbf{q}} \approx \mathbf{0}$ and velocities $\dot{\mathbf{q}} \approx \mathbf{0}$. The robot dynamic model is simplified as follows

$$\tau \approx \mathbf{g}(\mathbf{q}). \tag{10.6}$$

According to Eq. (10.6), the robot motors must above all compensate for the gravity effect. The model of gravitational effects $\hat{\mathbf{g}}(\mathbf{q})$ (the circumflex denotes the robot model), which is a good approximation of the actual gravitational forces $\mathbf{g}(\mathbf{q})$, can be implemented in the control algorithm shown in Fig. 10.4. The PD controller, shown in Fig. 10.2, was upgraded with an additional control loop, which calculates the gravitational forces from the actual robot position and directly adds them to the controller output. The control algorithm shown in Fig. 10.4 can be written as follows

$$\mathbf{u} = \mathbf{K}_p(\mathbf{q}_r - \mathbf{q}) - \mathbf{K}_d\dot{\mathbf{q}} + \hat{\mathbf{g}}(\mathbf{q}). \tag{10.7}$$

By introducing gravity compensation, the burden of reducing the errors caused by gravity is taken away from the PD controller. In this way the errors in trajectory tracking are significantly reduced.

10.1.3 Control of the Robot Based on Inverse Dynamics

When studying the PD controller with gravity compensation, we investigated the robot dynamic model in order to improve the efficiency of the control method. With the control method based on inverse dynamics, this concept will be further upgraded. From the equations describing the dynamic behavior of a two-segment robot manipulator (5.56), we can clearly observe that the robot model is nonlinear. A linear controller, such as the PD controller, is therefore not the best choice.

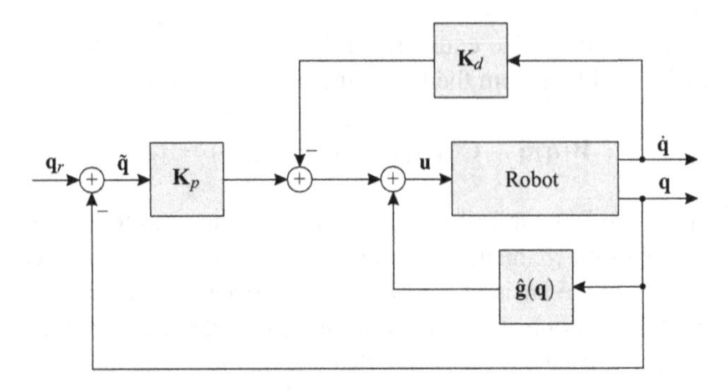

Fig. 10.4 PD control with gravity compensation

We shall derive the new control scheme from the robot dynamic model described by Eq. (10.5). Let us assume that the torques τ, generated by the motors, are equal to the control outputs **u**. Equation (10.5) can be rewritten

$$\mathbf{B}(\mathbf{q})\ddot{\mathbf{q}} + \mathbf{C}(\mathbf{q}, \dot{\mathbf{q}})\dot{\mathbf{q}} + \mathbf{F}_v\dot{\mathbf{q}} + \mathbf{g}(\mathbf{q}) = \mathbf{u}. \tag{10.8}$$

In the next step we will determine the direct robot dynamic model, which describes robot motions under the influence of the given joint torques. First we express the acceleration $\ddot{\mathbf{q}}$ from Eq. (10.8)

$$\ddot{\mathbf{q}} = \mathbf{B}^{-1}(\mathbf{q})\left(\mathbf{u} - (\mathbf{C}(\mathbf{q}, \dot{\mathbf{q}})\dot{\mathbf{q}} + \mathbf{F}_v\dot{\mathbf{q}} + \mathbf{g}(\mathbf{q}))\right). \tag{10.9}$$

By integrating the acceleration, while taking into account the initial velocity value, the velocity of robot motion is obtained. By integrating the velocity, while taking into account the initial position, we calculate the actual positions in the robot joints. The direct dynamic model of a robot mechanism is shown in Fig. 10.5.

In order to simplify the dynamic equations, we shall define a new variable $\mathbf{n}(\mathbf{q}, \dot{\mathbf{q}})$, comprising all dynamic components except the inertial component

$$\mathbf{n}(\mathbf{q}, \dot{\mathbf{q}}) = \mathbf{C}(\mathbf{q}, \dot{\mathbf{q}})\dot{\mathbf{q}} + \mathbf{F}_v\dot{\mathbf{q}} + \mathbf{g}(\mathbf{q}). \tag{10.10}$$

The robot dynamic model can be described with the following shorter equation

$$\mathbf{B}(\mathbf{q})\ddot{\mathbf{q}} + \mathbf{n}(\mathbf{q}, \dot{\mathbf{q}}) = \tau. \tag{10.11}$$

In the same way Eq. (10.9) can also be written in a shorter form

$$\ddot{\mathbf{q}} = \mathbf{B}^{-1}(\mathbf{q})\left(\mathbf{u} - \mathbf{n}(\mathbf{q}, \dot{\mathbf{q}})\right). \tag{10.12}$$

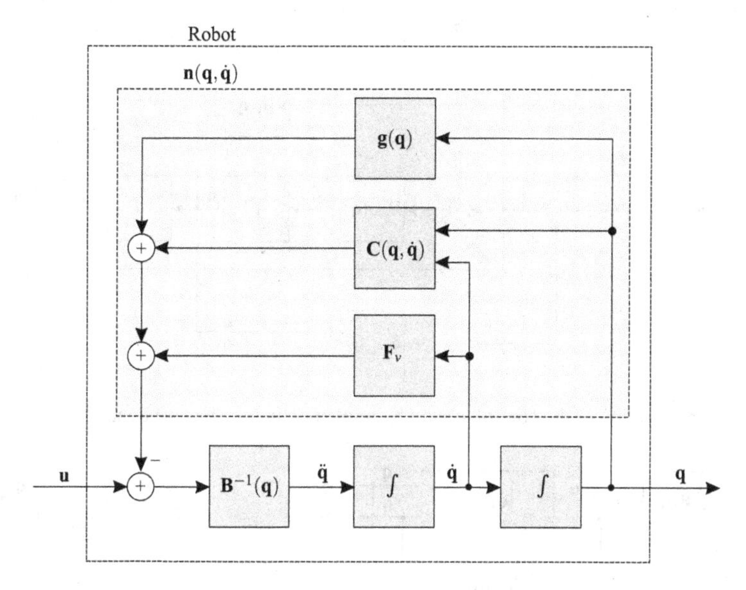

Fig. 10.5 The direct dynamic model of a robot mechanism

Let us assume that the robot dynamic model is known. The inertial matrix $\hat{\mathbf{B}}(\mathbf{q})$ is an approximation of the real values $\mathbf{B}(\mathbf{q})$, while $\hat{\mathbf{n}}(\mathbf{q}, \dot{\mathbf{q}})$ represents an approximation of $\mathbf{n}(\mathbf{q}, \dot{\mathbf{q}})$, as follows

$$\hat{\mathbf{n}}(\mathbf{q}, \dot{\mathbf{q}}) = \hat{\mathbf{C}}(\mathbf{q}, \dot{\mathbf{q}})\dot{\mathbf{q}} + \hat{\mathbf{F}}_v\dot{\mathbf{q}} + \hat{\mathbf{g}}(\mathbf{q}). \tag{10.13}$$

The controller output $\mathbf{u}$ is determined by the following equation

$$\mathbf{u} = \hat{\mathbf{B}}(\mathbf{q})\mathbf{y} + \hat{\mathbf{n}}(\mathbf{q}, \dot{\mathbf{q}}), \tag{10.14}$$

where the approximate inverse dynamic model of the robot was used. The system, combining Eqs. (10.12) and (10.14), is shown in Fig. 10.6.

Let us assume the equivalence $\hat{\mathbf{B}}(\mathbf{q}) = \mathbf{B}(\mathbf{q})$ and $\hat{\mathbf{n}}(\mathbf{q}, \dot{\mathbf{q}}) = \mathbf{n}(\mathbf{q}, \dot{\mathbf{q}})$. In Fig. 10.6 we observe that the signals $\hat{\mathbf{n}}(\mathbf{q}, \dot{\mathbf{q}})$ and $\mathbf{n}(\mathbf{q}, \dot{\mathbf{q}})$ subtract, as one is presented with a positive and the other with a negative sign. In a similar way, the product of matrices $\hat{\mathbf{B}}(\mathbf{q})$ and $\mathbf{B}^{-1}(\mathbf{q})$ results in a unit matrix, which can be omitted. The simplified system is shown in Fig. 10.7. By implementing the inverse dynamics (10.14), the control system is linearized, as there are only two integrators between the input $\mathbf{y}$ and the output $\mathbf{q}$. The system is not only linear, but is also decoupled (e.g. the first element of the vector $\mathbf{y}$ only influences the first element of the position vector $\mathbf{q}$). From Fig. 10.7 it is also not difficult to realize that the variable $\mathbf{y}$ has the characteristics of acceleration, thus

$$\mathbf{y} = \ddot{\mathbf{q}}. \tag{10.15}$$

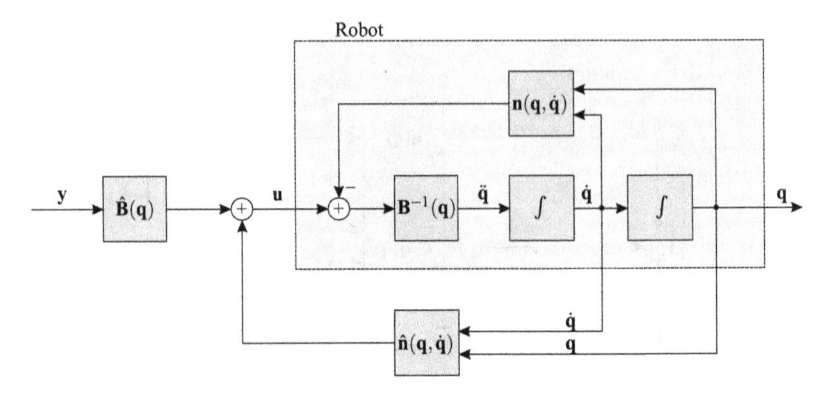

Fig. 10.6 Linearization of the control system by implementing the inverse dynamic model

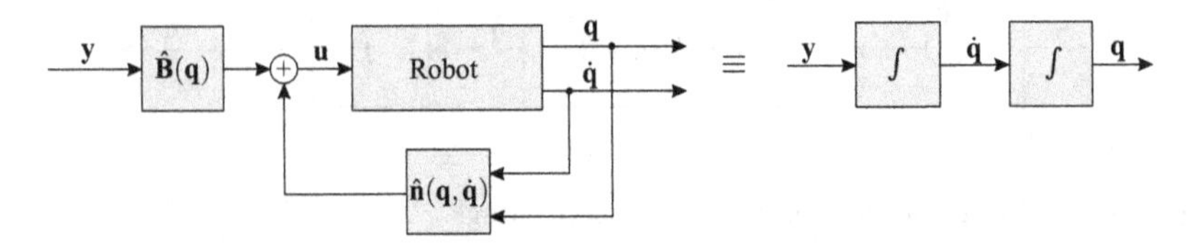

Fig. 10.7 The linearized system

In an ideal case, it would suffice to determine the desired joint accelerations as the second derivatives of the desired joint positions and the control system will track the prescribed joint trajectories. As we never have a fully accurate dynamic model of the robot, a difference will always occur between the desired and the actual joint positions and will increase with time. The positional error is defined by

$$\tilde{\mathbf{q}} = \mathbf{q}_r - \mathbf{q}, \tag{10.16}$$

where $\mathbf{q}_r$ represents the desired robot position. In a similar way also the velocity error can be defined as the difference between the desired and the actual velocity

$$\dot{\tilde{\mathbf{q}}} = \dot{\mathbf{q}}_r - \dot{\mathbf{q}}. \tag{10.17}$$

The vector $\mathbf{y}$, having the acceleration characteristics, can be now written as

$$\mathbf{y} = \ddot{\mathbf{q}}_r + \mathbf{K}_p(\mathbf{q}_r - \mathbf{q}) + \mathbf{K}_d(\dot{\mathbf{q}}_r - \dot{\mathbf{q}}). \tag{10.18}$$

It consists of the reference acceleration $\ddot{\mathbf{q}}_r$ and two contributing signals which depend on the errors of position and velocity. These two signals suppress the error arising because of the imperfectly modeled dynamics. The complete control scheme is shown in Fig. 10.8.

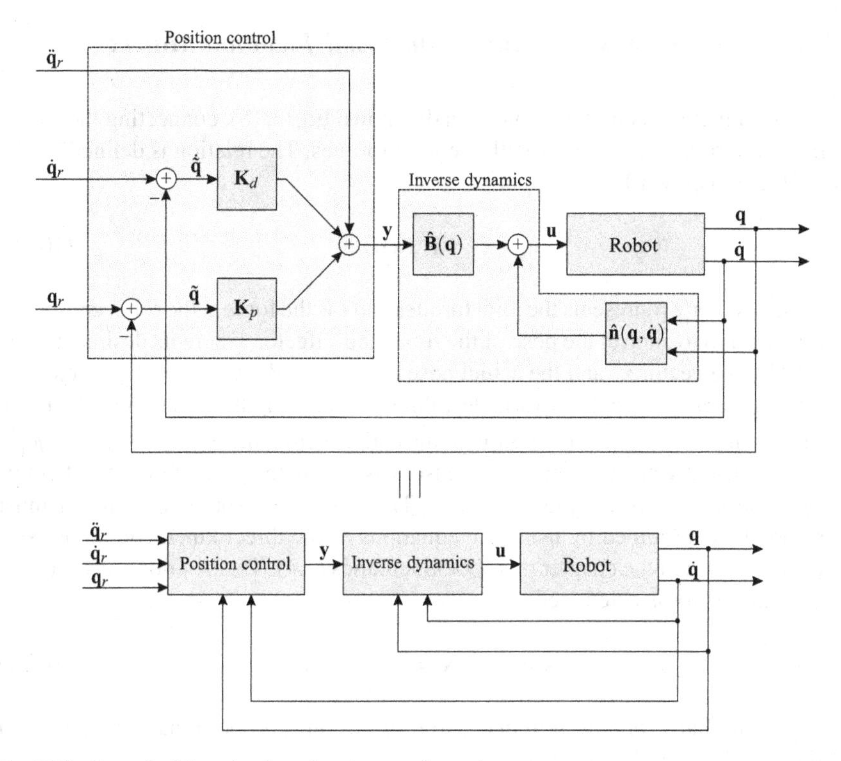

Fig. 10.8 Control of the robot based on inverse dynamics

By considering Eq. (10.18) and the equality $\mathbf{y} = \ddot{\mathbf{q}}$, the differential equation describing the robot dynamics can be written as

$$\ddot{\tilde{\mathbf{q}}} + \mathbf{K}_d\dot{\tilde{\mathbf{q}}} + \mathbf{K}_p\tilde{\mathbf{q}} = \mathbf{0}, \tag{10.19}$$

where the acceleration error $\ddot{\tilde{\mathbf{q}}} = \ddot{\mathbf{q}}_r - \ddot{\mathbf{q}}$ was introduced. The differential Eq. (10.19) describes the time dependence of the control error as it approaches zero. The dynamics of the response is determined by the gains $\mathbf{K}_p$ and $\mathbf{K}_d$.

10.2 Control of the Robot in External Coordinates

All the control schemes studied up to now were based on control of the internal coordinates (i.e., joint positions). The desired positions, velocities and accelerations were determined by the robot joint variables. Usually we are more interested in the motion of the robot end-effector than in the displacements of particular robot joints. At the tip of the robot, different tools are attached to accomplish various robot tasks. In the further text we shall focus on the robot control in the external coordinates.

10.2.1 Control Based on the Transposed Jacobian Matrix

The control method is based on the already known Eq. (5.18), connecting the forces acting at the robot end-effector with the joint torques. The relation is defined by the use of the transposed Jacobian matrix

$$\tau = \mathbf{J}^T(\mathbf{q})\mathbf{f}, \tag{10.20}$$

where the vector τ represents the joint torques and $\mathbf{f}$ is the force at the robot end-point.

It is our aim to control the pose of the robot end-effector, where its desired pose is defined by the vector $\mathbf{x}_r$ and the actual pose is given by the vector $\mathbf{x}$. The vectors $\mathbf{x}_r$ and $\mathbf{x}$ in general comprise six variables, three determining the position of the robot end-point and three for the orientation of the end-effector, thus $\mathbf{x} = \begin{bmatrix} x & y & z & \varphi & \vartheta & \psi \end{bmatrix}^T$. Robots are usually not equipped with sensors assessing the pose of the end-effector; robot sensors measure the joint variables. The pose of the robot end-effector must therefore, be determined by using the equations of the direct kinematic model $\mathbf{x} = \mathbf{k}(\mathbf{q})$, introduced in the chapter on robot kinematics (5.4). The positional error of the robot end-effector is calculated as

$$\tilde{\mathbf{x}} = \mathbf{x}_r - \mathbf{x} = \mathbf{x}_r - \mathbf{k}(\mathbf{q}). \tag{10.21}$$

The positional error must be reduced to zero. A simple proportional control system with the gain matrix $\mathbf{K}_p$ is introduced

$$\mathbf{f} = \mathbf{K}_p\tilde{\mathbf{x}}. \tag{10.22}$$

When analyzing Eq. (10.22) more closely, we find that it reminds us of the equation describing the behavior of a spring (in external coordinates), where the force is proportional to the spring elongation. This consideration helps us to explain the introduced control principle. Let as imagine that there are six springs virtually attached to the robot end-effector, one spring for each degree of freedom (three for position and three for orientation). When the robot moves away from the desired pose, the springs are elongated and pull the robot end-effector into the desired pose with the force proportional to the positional error. The force $\mathbf{f}$ therefore pushes the robot end-effector towards the desired pose. As the robot displacement can only be produced by the motors in the joints, the variables controlling the motors must be calculated from the force $\mathbf{f}$. This calculation is performed by the help of the transposed Jacobian matrix as shown in Eq. (10.20)

$$\mathbf{u} = \mathbf{J}^T(\mathbf{q})\mathbf{f}. \tag{10.23}$$

The vector $\mathbf{u}$ represents the desired joint torques. The control method based on the transposed Jacobian matrix is shown in Fig. 10.9.

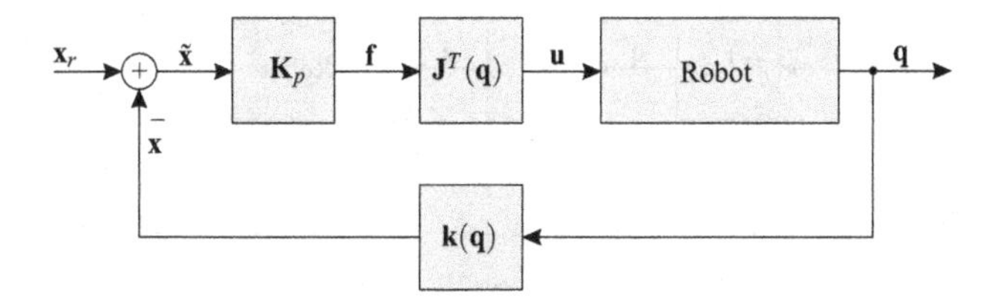

Fig. 10.9 Control based on the transposed Jacobian matrix

10.2.2 Control Based on the Inverse Jacobian Matrix

The control method is based on the relation between the joint velocities and the velocities of the robot end-point (5.10), which is given by the Jacobian matrix. In Eq. (5.10) we emphasize the time derivatives of external coordinates $\mathbf{x}$ and internal coordinates $\mathbf{q}$

$$\dot{\mathbf{x}} = \mathbf{J}(\mathbf{q})\dot{\mathbf{q}} \quad \Leftrightarrow \quad \frac{d\mathbf{x}}{dt} = \mathbf{J}(\mathbf{q})\frac{d\mathbf{q}}{dt}. \tag{10.24}$$

As dt appears in the denominator on both sides of Eq. (10.24), it can be omitted. In this way we obtain the relation between changes of the internal coordinates and changes of the pose of the robot end-point

$$d\mathbf{x} = \mathbf{J}(\mathbf{q})d\mathbf{q}. \tag{10.25}$$

Equation (10.25) is valid only for small displacements.

As with the previously studied control method, based on the transposed Jacobian matrix, we can also in this case first calculate the error of the pose of the robot end-point by using Eq. (10.21). When the error in the pose is small, we can calculate the positional error in the internal coordinates by the inverse relation (10.25)

$$\tilde{\mathbf{q}} = \mathbf{J}^{-1}(\mathbf{q})\tilde{\mathbf{x}}. \tag{10.26}$$

In this way the control method is translated to the known method of robot control in the internal coordinates. In the simplest example, based on the proportional controller, we can write

$$\mathbf{u} = \mathbf{K}_p\tilde{\mathbf{q}}. \tag{10.27}$$

The equation describes the behavior of a spring (in internal coordinates). The control method, based on the inverse Jacobian matrix, is shown in Fig. 10.10.

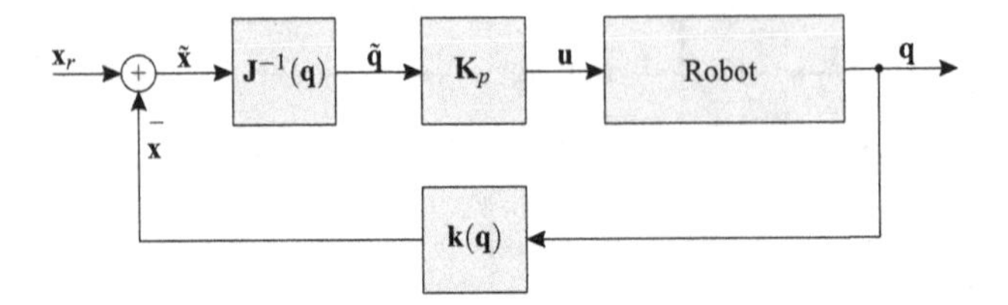

Fig. 10.10 Control based on the inverse Jacobian matrix

10.2.3 PD Control of Position with Gravity Compensation

The PD control of position with gravity compensation was already studied in detail for the internal coordinates. Now we shall derive the analogue control algorithm in the external coordinates. The starting point will be Eq. (10.21), expressing the error of the pose of the end-effector. The velocity of the robot end-point is calculated with the help of the Jacobian matrix from the joint velocities

$$\dot{\mathbf{x}} = \mathbf{J}(\mathbf{q})\dot{\mathbf{q}}. \tag{10.28}$$

The equation describing the PD controller in external coordinates is analogous to that written in the internal coordinates (10.2)

$$\mathbf{f} = \mathbf{K}_p\tilde{\mathbf{x}} - \mathbf{K}_d\dot{\mathbf{x}}. \tag{10.29}$$

In Eq. (10.29), the pose error is multiplied by the matrix of the positional gains $\mathbf{K}_p$, while the velocity error is multiplied by the matrix $\mathbf{K}_d$. The negative sign of the velocity error introduces damping into the system. The joint torques are calculated from the force $\mathbf{f}$, acting at the tip of the robot, with the help of the transposed Jacobian matrix (in a similar way as in Eq. (10.23)) and by adding the component compensating gravity (as in Eq. (10.7)). The control algorithm is written as

$$\mathbf{u} = \mathbf{J}^T(\mathbf{q})\mathbf{f} + \hat{\mathbf{g}}(\mathbf{q}). \tag{10.30}$$

The complete control scheme is shown in Fig. 10.11.

10.2.4 Control of the Robot Based on Inverse Dynamics

In the chapter on the control of robots in the internal coordinates, the following controller based on inverse dynamics was introduced

$$\mathbf{u} = \hat{\mathbf{B}}(\mathbf{q})\mathbf{y} + \hat{\mathbf{n}}(\mathbf{q}, \dot{\mathbf{q}}). \tag{10.31}$$

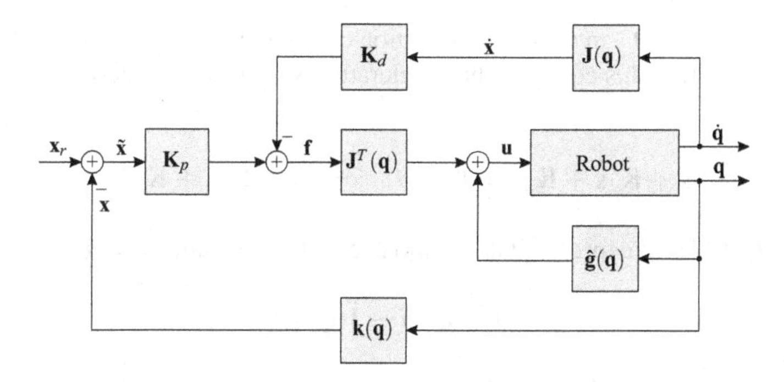

Fig. 10.11 PD control with gravity compensation in external coordinates

We also learned that the vector $\mathbf{y}$ has the characteristics of acceleration

$$\mathbf{y} = \ddot{\mathbf{q}}, \tag{10.32}$$

which was determined in such a way, that the robot tracked the desired trajectory expressed in the internal coordinates. As it is our aim to develop a control method in the external coordinates, the $\mathbf{y}$ signal must be adequately adapted. Equation (10.31), linearizing the system, remains unchanged.

We shall again start from the equation relating the joint velocities to the robot end-effector velocities

$$\dot{\mathbf{x}} = \mathbf{J}(\mathbf{q})\dot{\mathbf{q}}. \tag{10.33}$$

By calculating the time derivative of Eq. (10.33), we obtain

$$\ddot{\mathbf{x}} = \mathbf{J}(\mathbf{q})\ddot{\mathbf{q}} + \dot{\mathbf{J}}(\mathbf{q}, \dot{\mathbf{q}})\dot{\mathbf{q}}. \tag{10.34}$$

The error of the pose of the robot end-effector is determined as the difference between its desired and its actual pose

$$\tilde{\mathbf{x}} = \mathbf{x}_r - \mathbf{x} = \mathbf{x}_r - \mathbf{k}(\mathbf{q}). \tag{10.35}$$

In a similar way the velocity error of the robot end-effector is determined

$$\dot{\tilde{\mathbf{x}}} = \dot{\mathbf{x}}_r - \dot{\mathbf{x}} = \dot{\mathbf{x}}_r - \mathbf{J}(\mathbf{q})\dot{\mathbf{q}}. \tag{10.36}$$

The acceleration error is the difference between the desired and the actual acceleration

$$\ddot{\tilde{\mathbf{x}}} = \ddot{\mathbf{x}}_r - \ddot{\mathbf{x}}. \tag{10.37}$$

When developing the inverse dynamics based controller in the internal coordinates, Eq. (10.19) was derived describing the dynamics of the control error in the form

$\ddot{\tilde{\mathbf{q}}} + \mathbf{K}_d\dot{\tilde{\mathbf{q}}} + \mathbf{K}_p\tilde{\mathbf{q}} = \mathbf{0}$. An analogous equation can be written for the error of the end-effector pose. From this equation the acceleration $\ddot{\mathbf{x}}$ of the robot end-effector can be expressed

$$\ddot{\tilde{\mathbf{x}}} + \mathbf{K}_d\dot{\tilde{\mathbf{x}}} + \mathbf{K}_p\tilde{\mathbf{x}} = \mathbf{0} \quad \Rightarrow \quad \ddot{\mathbf{x}} = \ddot{\mathbf{x}}_r + \mathbf{K}_d\dot{\tilde{\mathbf{x}}} + \mathbf{K}_p\tilde{\mathbf{x}}. \tag{10.38}$$

From Eq. (10.34) we express $\ddot{\mathbf{q}}$ taking into account the equality $\mathbf{y} = \ddot{\mathbf{q}}$

$$\mathbf{y} = \mathbf{J}^{-1}(\mathbf{q})\left(\ddot{\mathbf{x}} - \dot{\mathbf{J}}(\mathbf{q}, \dot{\mathbf{q}})\dot{\mathbf{q}}\right). \tag{10.39}$$

By replacing $\ddot{\mathbf{x}}$ in Eq. (10.39) with expression (10.38), the control algorithm based on inverse dynamics in the external coordinates is obtained

$$\mathbf{y} = \mathbf{J}^{-1}(\mathbf{q})\left(\ddot{\mathbf{x}}_r + \mathbf{K}_d\dot{\tilde{\mathbf{x}}} + \mathbf{K}_p\tilde{\mathbf{x}} - \dot{\mathbf{J}}(\mathbf{q}, \dot{\mathbf{q}})\dot{\mathbf{q}}\right). \tag{10.40}$$

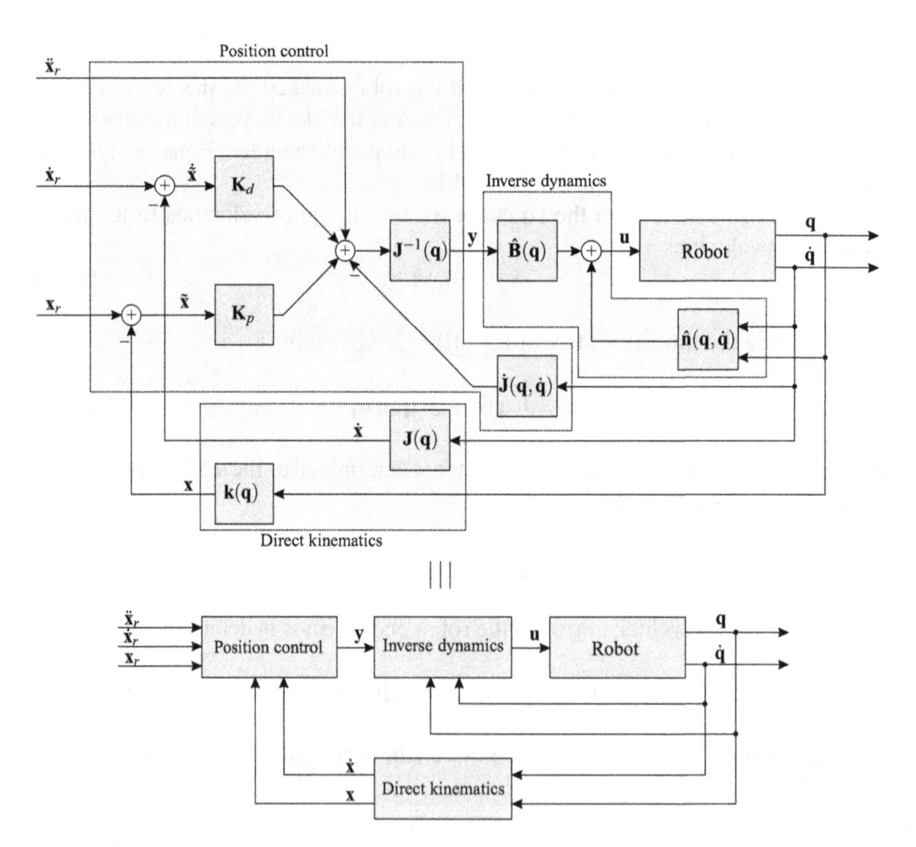

Fig. 10.12 Robot control based on inverse dynamics in external coordinates

The control scheme encompassing the linearization of the system based on inverse dynamics (10.31) and the closed loop control (10.40) is shown in Fig. 10.12.

10.3 Control of the Contact Force

The control of position is sufficient when a robot manipulator follows a trajectory in free space. When contact occurs between the robot end-effector and the environment, position control is not an appropriate approach. Let us imagine a robot manipulator cleaning a window with a sponge. As the sponge is very compliant, it is possible to control the force between the robot and window by controlling the position between the robot gripper and the window. If the sponge is sufficiently compliant and when we know the position of the window accurately enough, the robot will appropriately accomplish the task.

If the compliance of the robot tool or its environment is smaller, then it is not so simple to execute the tasks which require contact between the robot and its environment. Let us now imagine a robot scraping paint from a glassy surface while using a stiff tool. Any uncertainty in the position of the glassy surface or malfunction of the robot control system will prevent satisfactory execution of the task; either the glass will break, or the robot will wave uselessly in thin air.

In both robot tasks, i.e. cleaning a window or scraping a smooth surface, it is more reasonable that instead of position of the glassy surface we determine the force that the robot should exert on the environment. Most of the modern industrial robots carry out relatively simple tasks, such as spot welding, spray painting, and various point-to-point operations. Several robot applications, however, require control of the contact force. A characteristic example is grinding or a similar robot machining task. An important area of industrial robotics is also robot assembly, where several component parts are to be assembled. In such robot tasks, sensing and controlling the forces is of utmost importance.

Accurate operation of a robot manipulator in an uncertain, non-structured, and changeable environment is required for efficient use of robots in an assembly task. Here, several component parts must be brought together with high accuracy. Measurement and control of the contact forces enable the required positional accuracy of the robot manipulator to be reached. As relative measurements are used in robot force control, the absolute errors in positioning of either the manipulator or the object are not as critical as in robot position control. When dealing with stiff objects, already small changes in position produce large contact forces. Measurement and control of those forces can lead to significantly higher positional accuracy of robot movement.

When a robot is exerting force on the environment, we deal with two types of robot tasks. In the first case we would like the robot end-effector to be brought into a desired pose while the robot is in contact with the environment. This is the case of robot assembly. A characteristic example is that of inserting a peg into a hole. The robot movement must be of such nature that the contact force is reduced to zero or to a minimal value allowed. In the second type of robot task, we require from the robot end-effector to exert a predetermined force on the environment. This is the example of robot grinding. Here, the robot movement depends on the difference between the desired and the actual measured contact force.

The robot force control method will be based on control of the robot using inverse dynamics. Because of the interaction of the robot with the environment, an additional component, representing the contact force $\mathbf{f}$, appears in the inverse dynamic model. As the forces acting at the robot end-effector are transformed into the joint torques by the use of the transposed Jacobian matrix (5.18), we can write the robot dynamic model in the following form

$$\mathbf{B}(\mathbf{q})\ddot{\mathbf{q}} + \mathbf{C}(\mathbf{q}, \dot{\mathbf{q}})\dot{\mathbf{q}} + \mathbf{F}_v\dot{\mathbf{q}} + \mathbf{g}(\mathbf{q}) = \tau - \mathbf{J}^T(\mathbf{q})\mathbf{f}. \tag{10.41}$$

On the right hand side of the Eq. (10.5) we added the component $-\mathbf{J}^T(\mathbf{q})\mathbf{f}$ representing the force of interaction with the environment. It can be seen that the force $\mathbf{f}$ acts through the transposed Jacobian matrix in a similar way as the joint torques (i.e., it tries to produce robot motion). The model (10.41) can be rewritten in a shorter form by introducing

$$\mathbf{n}(\mathbf{q}, \dot{\mathbf{q}}) = \mathbf{C}(\mathbf{q}, \dot{\mathbf{q}})\dot{\mathbf{q}} + \mathbf{F}\dot{\mathbf{q}} + \mathbf{g}(\mathbf{q}), \tag{10.42}$$

which gives us the following dynamic model of a robot in contact with its environment

$$\mathbf{B}(\mathbf{q})\ddot{\mathbf{q}} + \mathbf{n}(\mathbf{q}, \dot{\mathbf{q}}) = \tau - \mathbf{J}^T(\mathbf{q})\mathbf{f}. \tag{10.43}$$

10.3.1 Linearization of a Robot System Through Inverse Dynamics

Let us denote the control output, representing the desired actuation torques in the robot joints, by the vector $\mathbf{u}$. Equation (10.43) can be written as follows

$$\mathbf{B}(\mathbf{q})\ddot{\mathbf{q}} + \mathbf{n}(\mathbf{q}, \dot{\mathbf{q}}) + \mathbf{J}^T(\mathbf{q})\mathbf{f} = \mathbf{u}. \tag{10.44}$$

From Eq. (10.44) we express the direct dynamic model

$$\ddot{\mathbf{q}} = \mathbf{B}^{-1}(\mathbf{q})\left(\mathbf{u} - \mathbf{n}(\mathbf{q}, \dot{\mathbf{q}}) - \mathbf{J}^T(\mathbf{q})\mathbf{f})\right). \tag{10.45}$$

Equation (10.45) describes the response of the robot system to the control input $\mathbf{u}$. By integrating the acceleration, while taking into account the initial velocity value, the actual velocity of the robot motion is obtained. By integrating the velocity, while taking into the account the initial position, we calculate the actual positions in the robot joints. The described model is represented by the block *Robot* in Fig. 10.13.

In a similar way as when developing the control method based on inverse dynamics, we will linearize the system by including the inverse dynamic model into the closed loop

$$\mathbf{u} = \hat{\mathbf{B}}(\mathbf{q})\mathbf{y} + \hat{\mathbf{n}}(\mathbf{q}, \dot{\mathbf{q}}) + \mathbf{J}^T(\mathbf{q})\mathbf{f}, \tag{10.46}$$

The use of circumflex denotes the estimated parameters of the robot system. The difference between Eqs. (10.46) and (10.14), representing the control based on inverse dynamics in internal coordinates, is the component $\mathbf{J}^T(\mathbf{q})\mathbf{f}$, compensating the influence of external forces on the robot mechanism. The control scheme, combining Eqs. (10.45) and (10.46), is shown in Fig. 10.13. Assuming that the estimated parameters are equal to the actual robot parameters, it can be observed, that by introducing the closed loop (10.46), the system is linearized because there are only two integrators between the input $\mathbf{y}$ and the output $\mathbf{q}$, as already demonstrated in Fig. 10.7.

10.3.2 Force Control

After linearizing the control system, the input vector $\mathbf{y}$ must be determined. The force control will be translated to control of the pose of the end-effector. This can be, in a simplified way, explained with the following reasoning: if we wish the robot to increase the force exerted on the environment, the robot end-effector must be displaced in the direction of the action of the force. Now we can use the control system which was developed to control the robot in the external coordinates (10.40). The control scheme of the robot end-effector including the linearization, while taking into account the contact force, is shown in Fig. 10.14.

Up to this point we mainly summarized the knowledge of the pose control of the robot end-effector as explained in the previous chapters. In the next step we will determine the desired pose, velocity and acceleration of the robot end-effector, on the basis of the force measured between the robot end-point and its environment.

Let us assume that we wish to control a constant desired force $\mathbf{f}_r$. With the force wrist sensor, the contact force $\mathbf{f}$ is measured. The difference between the desired and measured force represents the force error

$$\tilde{\mathbf{f}} = \mathbf{f}_r - \mathbf{f}. \tag{10.47}$$

The desired robot motion will be calculated based on the assumption that the force $\tilde{\mathbf{f}}$ must displace a virtual object with inertia $\mathbf{B}_c$ and damping $\mathbf{F}_c$. In our case the virtual object is in fact the robot end-effector. For easier understanding, let us consider a system with only one degree of freedom. When a force acts on such a system, an accelerated movement will start. The movement will be determined by the force, the mass of the object and the damping. The robot end-effector therefore behaves as a system consisting of a mass and a damper, which are under the influence of the

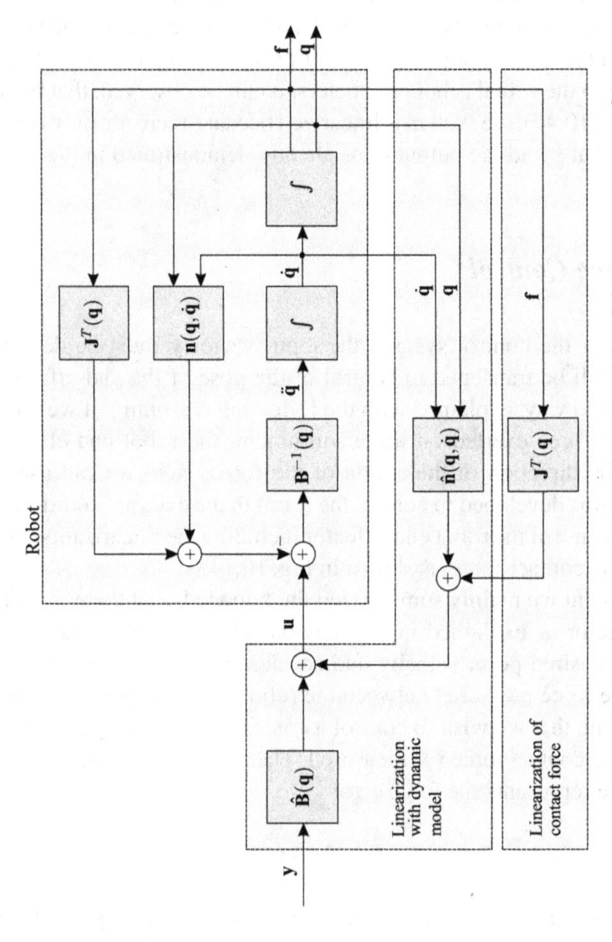

Fig. 10.13 Linearization of the control system by implementing the inverse dynamic model and the measured contact force

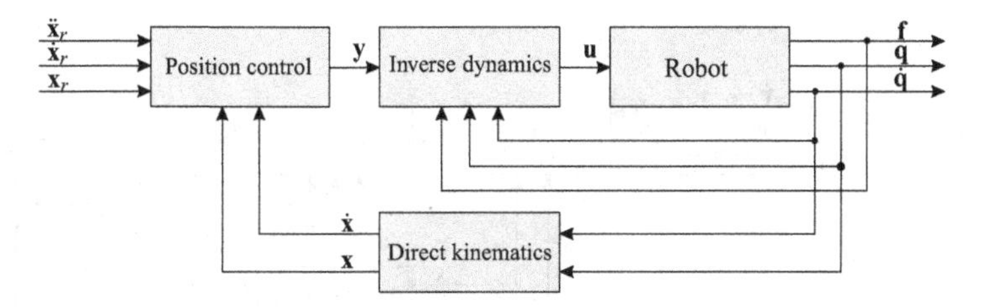

Fig. 10.14 Robot control based on inverse dynamics in external coordinates including the contact force

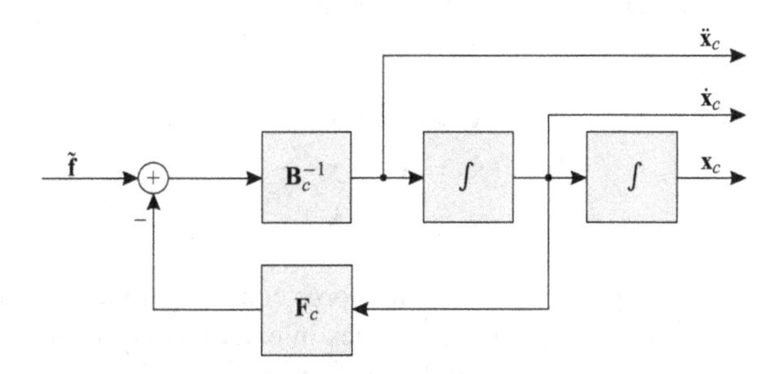

Fig. 10.15 Force control translated into control of the pose of robot end-effector

force $\tilde{\mathbf{f}}$. For more degrees of freedom we can write the following differential equation describing the movement of the object

$$\tilde{\mathbf{f}} = \mathbf{B}_c\ddot{\mathbf{x}}_c + \mathbf{F}_c\dot{\mathbf{x}}_c. \tag{10.48}$$

The matrices $\mathbf{B}_c$ and $\mathbf{F}_c$ determine the movement of the object under the influence of the force $\tilde{\mathbf{f}}$. From Eq. (10.48) the acceleration of the virtual object can be calculated

$$\ddot{\mathbf{x}}_c = \mathbf{B}_c^{-1}\left(\tilde{\mathbf{f}} - \mathbf{F}_c\dot{\mathbf{x}}_c\right). \tag{10.49}$$

By integrating the Eq. (10.49), the velocities and the pose of the object are calculated, as shown in Fig. 10.15. In this way the reference pose $\mathbf{x}_c$, reference velocity $\dot{\mathbf{x}}_c$, and reference acceleration $\ddot{\mathbf{x}}_c$ are determined from the force error. The calculated variables are inputs to the control system, shown in Fig. 10.14. In this way the force control was translated into the already known robot control in external coordinates.

In order to also simultaneously control the pose of the robot end-effector, parallel composition is included. Parallel composition assumes that the reference control variables are obtained by summing the references for force control ($\mathbf{x}_c$, $\dot{\mathbf{x}}_c$, $\ddot{\mathbf{x}}_c$) and references for the pose control ($\mathbf{x}_d$, $\dot{\mathbf{x}}_d$, $\ddot{\mathbf{x}}_d$). The parallel composition is defined by equations

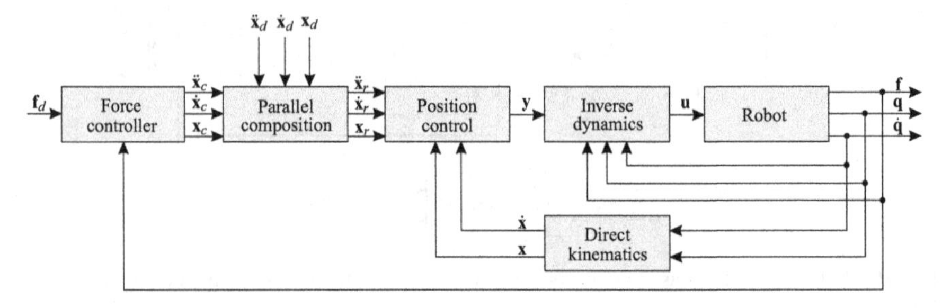

Fig. 10.16 Direct force control in the external coordinates

$$\begin{aligned} \mathbf{x}_r &= \mathbf{x}_d + \mathbf{x}_c \\ \dot{\mathbf{x}}_r &= \dot{\mathbf{x}}_d + \dot{\mathbf{x}}_c \\ \ddot{\mathbf{x}}_r &= \ddot{\mathbf{x}}_d + \ddot{\mathbf{x}}_c \end{aligned} \tag{10.50}$$

The control system incorporating the contact force control, parallel composition and control of the robot based on inverse dynamics in external coordinates is shown in Fig. 10.16. The force control is obtained by selecting

$$\begin{aligned} \mathbf{x}_r &= \mathbf{x}_c \\ \dot{\mathbf{x}}_r &= \dot{\mathbf{x}}_c \\ \ddot{\mathbf{x}}_r &= \ddot{\mathbf{x}}_c \end{aligned} \tag{10.51}$$

The described control method enables the control of force. However, it does not enable independent control of the pose of the robot end-effector as it is determined by the error in the force signal.

Chapter 11
Robot Environment

This chapter will illustrate robot environments, exemplified by product assembly processes where robots are a part of a production line or as completely independent units. The example can be easily replicated also to other tasks, such as product inspection and testing, welding, painting, pick and place operations etc.

As a matter of fact, robots represent an ideal solution for many industrial safety and health problems, mainly because they are capable of performing hard and fatiguing tasks in a dangerous environment. Welding and painting robots enable human workers to avoid toxic fumes and vapors. Robots also load power presses, which were frequent causes of injuries to workers in the past. Robots work in foundries and radioactive environments. With the increasing number of robots in industrial processes, there is, however, an increased danger introduced by the robots themselves. Thus, considering safety is of utmost importance when designing a robotic working cell.

11.1 Robot Safety

Industrial robots are strong devices which move quickly in their workspace. An accident in most cases occurs only when a human worker enters the robot workspace. A person steps into the robot vicinity either accidentally or even without knowing or with the aim of robot reprogramming or maintenance. It is often difficult for a human operator to judge what will be the robot's next move. Particularly dangerous are the unexpected robot motions, which are the consequence either of a robot failure or of a programming error. Many governmental organizations and large companies, together with robot producers, have developed safety standards. The approaches assuring safe cooperation of human workers and industrial robots can be divided into three major groups: (1) robot safety features, (2) robot workspace safeguards, and (3) personnel training and supervision.

M. Mihelj et al., *Robotics*, https://doi.org/10.1007/978-3-319-72911-4_11

Today's robots have safety features to a large extent already built-in for all three modes of operation: normal work, programming and maintenance. Fault avoidance features increase robot reliability and safety. Such a feature, for example, prevents the robot from reaching into the press before it is open. The safety features built into the robot control unit usually enable synchronization between the robot and other machines in the robot environment. Checking the signals, indicating when a device is ready to take an active part in the robot cell, must be part of safe robot programming. The use of reliable sensors plays an important role when checking the status of machines in the robot working area. Important safety features of any robot system are also software and electric stops.

When programming or teaching a robot, the human operator must be in the robot working area. In the programming phase the velocity of the robot motions must be considerably lower than during normal work. The speed of the robot must be reduced to such a value that the human operator can avoid unexpected robot motions. The recommended maximal velocity of the robot, when there is a human worker inside the workspace, is 0.25 m/s.

The teach pendant unit can be a critical component in safe robot operation. Programming errors during teaching of a robot often cause unexpected robot motions. The design of a teach pendant unit can have a significant impact on safe operation. The use of joystick control was found safer than the use of control pushbuttons. The size of emergency pushbuttons also has an important influence on the human operator's reaction times.

Special safety features facilitate safe robot maintenance. Such a feature is, for example, the possibility of switching on the control system, while the robot arm is not powered. Another feature enables passive manual motion of the robot segments, while the robot actuators are switched off. Some robot features cause the robot to stop as soon as possible, while some allow the control system to execute the current command and stop afterwards.

Most robot accidents occur when persons intentionally or carelessly enter the robot working area. The robot workspace safeguards prevent such entrance into the robot cell. There are three major approaches to the robot workspace protection: (1) barriers and fences, (2) presence sensing, and (3) warning signs, signals and lights.

Most commonly metal barriers or fences are used to prevent unauthorized workers from entering the robot working area. The color of the fence plays an important role, efficiently warning non-informed personnel. The fences are also an adequate protection against various vehicles that are used for transporting materials in the production hall. Safe opening of the gates, which enable entrance into the fenced-off area, must also be provided. A human operator can only enter after switching-off the robot system using a control panel outside the barriers. Well-designed safeguarding barriers may also protect bystanders from objects flying out from the robot's grasp.

Important safeguarding is provided by the devices detecting the presence of a person in the robot working area. These can be pressure-sensitive floor mats, light curtains, end-effector sensors, various ultrasound, capacitive, infrared or microwave sensors inside the robot cell and computer vision. Instrumented floor mats or light curtains can detect the entrance of a person into the robot working area. In such a case a warning signal is triggered and normal robot working can be stopped. The end-effector sensors detect the unexpected collisions with objects in the robot environment and cause an emergency stop. Contactless sensors and computer vision detect various intrusions into the robot working area.

Warning signs, signals and lights can to a large extent increase the safe operation of robot cells. These warning signs alert the operators to the presence of a hazardous situation. Instruction manuals and proper training are also important for effective use of warning signs. Such signs are more effective with people who unintentionally enter the robot working area, than with operators who are familiar with the operation of the robot cell. Experienced operators often neglect the warnings and intentionally enter the robot workspace without switching off the robot aiming to save some small amount of time. Such moves are often causes of accidents. False alarms may also reduce the effectiveness of warnings.

Selection of qualified workers, safety training and proper supervision are the pre-requisites for safe working with robots. Especially critical moments are startup and shutdown of a robot cell. Similarly, maintenance and programming of robots can be dangerous. Some robot applications (e.g. welding) include specific dangerous situations which must be well known to the workers. Those employed in the robot environment must satisfy both physical and mental requirements for their job. The selection of appropriate workers is an important first step. The second step, which is equally important, is extensive safety training. Satisfactory safety is only achieved with constant supervision of the employees. Additional training is an important component of the application of industrial robots. In the training courses the workers must be acquainted with the possible hazards and their severity. They must learn how to identify and avoid hazardous situations. Common mistakes that are causes of accidents should be explained in detail. Such training courses are usually prepared with the help of robot manufacturers.

It is expected that future robots will not work behind safety guards with locked doors or light barriers. Instead they will be working in close cooperation with humans which leads to the fundamental concern of how to ensure a safe human-robot physical interaction. The major progress is expected in the design of lightweight flexible robot segments, compliant joints, novel actuators and advanced control algorithms.

The robot installation can be as an individual robot cell or as a part of a larger industrial production line. Industrial robots are position controlled and often without sensors for sensing their surroundings. For this reason the robots must be isolated from human environment in case of improper activity of the robot or its peripheral parts, to prevent human injuries or collisions with other equipment in the robot working cell. The safety risk for each individual robot cell needs to be defined so that appropriate precautions can be taken. Improper robot behavior can be the result of robot system fault or human error, such as:

- unpredictable robot behavior because of a fault in the control system,
- cable connection fault because of robot movement,
- data transfer error producing unpredictable robot movement,
- robot tool fault, e.g. welding gun,

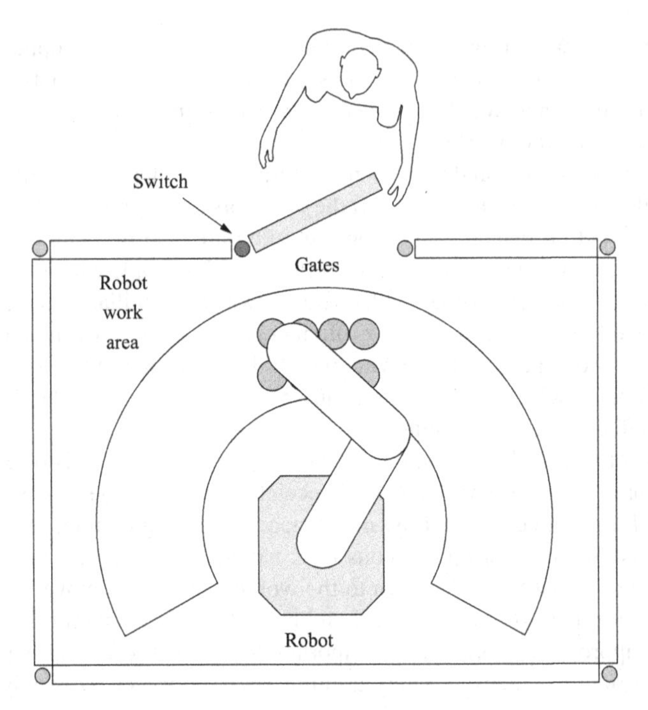

Fig. 11.1 Level 1: mechanical robot cell protection

- software errors,
- worn out robot mechanical components.

The potential dangers of system faults arising from these errors can be divided into three categories:

- **Risk of collision** is the possibility that the moving robot or the tool attached to the robot hits the operator.
- **Danger of pinching** is a situation where the robot, during the movement near the objects in the robot cell (e.g. transport mechanisms), squeezes the operator.
- **Other hazards** that are specific to each robot application, such as the risk of electrical shock, impact of welding arc, burns, toxic substances, radiation, excessive sound levels.

For all these reasons the robot safety demands can be split in three levels.

Level 1 is the level of protection of the entire robot cell. It is usually achieved with physical protection using a combination of mechanical fences, railings and gates (Fig. 11.1). Alongside physical protection also a human presence sensor (e.g. laser curtains) can be installed.

Level 2 includes a level of protection while an operator is in the working area of the robot. Normally, protection is performed by presence sensors. In contrast to the previous level, which is based mainly on mechanical protection, level 2 is based on the perception of the operators presence (Fig. 11.2).

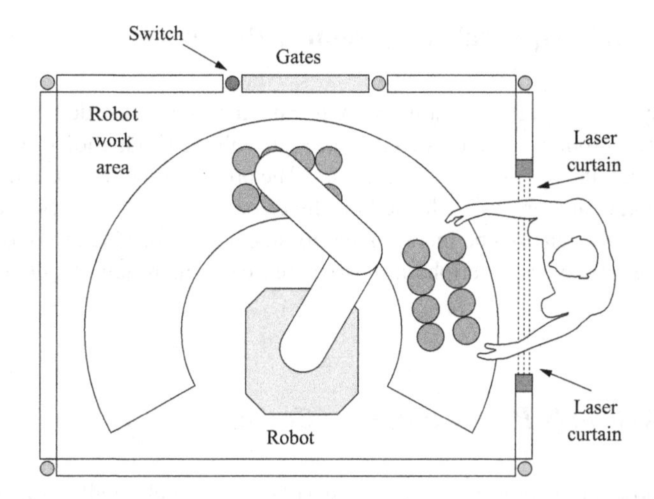

Fig. 11.2 Level 2: opto-electrical robot cell protection

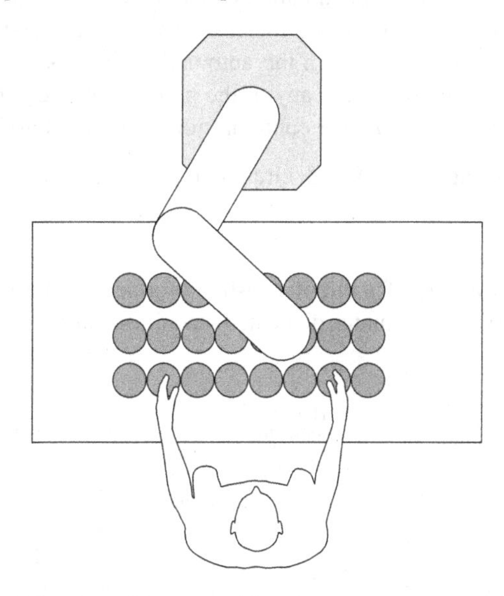

Fig. 11.3 Level 3: collaboration of human and the robot

Level 3 is the level of protection where people are in contact with the robot referred to as collaborative robots. Security at this level is carried out by detecting the presence of a human or obstacles nearby the robot or when the robot and the human are in collaboration (Fig. 11.3). In risk situations the robot system must slow down or stop. These systems incorporate sensors for human tracking, various force and torque sensors and contact or touch sensors. Collaborative robots are described in more details in Chap. 12.

11.2 Robot Peripherals in Assembly Processes

The robot systems installed in industry are usually a part of larger dedicated production lines. The production lines are used for high-volume production of parts where multiple processing operations are necessary. The production line is split in workstations where human workers, dedicated machines or robots perform necessary tasks. Other peripherals can also be incorporated to increase the production line capabilities. The properly selected peripherals also increase system reliability, flexibility and efficiency.

11.2.1 Assembly Production Line Configurations

Assembly production lines in industry consist of conveyor belts, pallets traveling with conveyor belts, vision systems, pneumatic cylinders, different sensors and robots or manipulators. The pallets provide the mean to index, locate and track individual manufactured parts traveling through the automation process. The robots provide flexibility and can be integrated into any of the production line configurations. The most usual assembly production line configurations with robot assistance are:

- In-line (direct, L-shaped, U-shaped, circular, rectangular),
- Rotary,
- Hybrid.

In Fig. 11.4 an example of a circular in-line production line is presented. Line workstations are served by humans, dedicated machines and robots. Parts for assem-

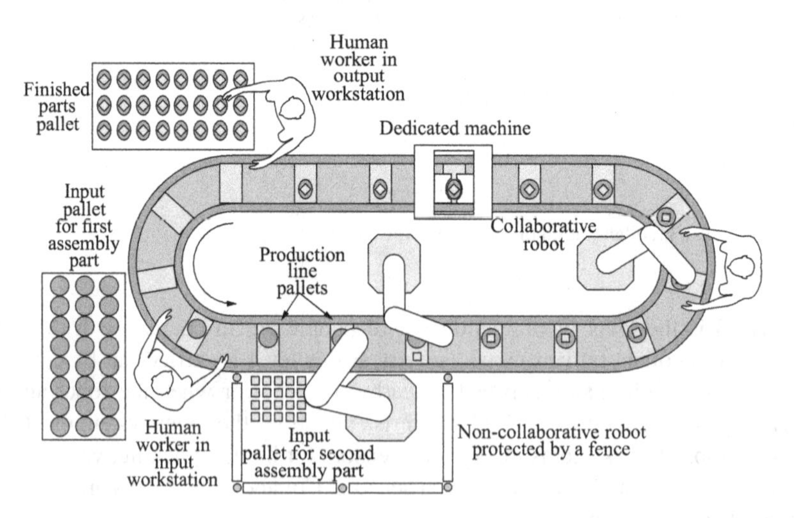

Fig. 11.4 Example of circular in-line assembly production line with human, machine and robot workstations

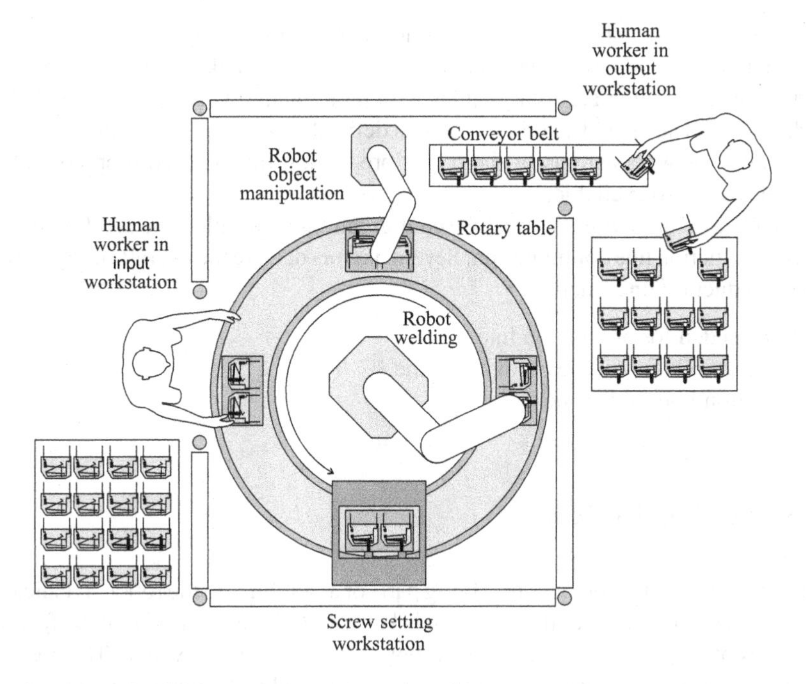

Fig. 11.5 Example of rotary table production line with human, machine and robot workstations

bly are manipulated by hand or by the robot and transferred among workstations by pallets along a conveyor belt. The distance between the pallets is not necessarily constant and their position is monitored by location sensors, usually capacitive or inductive presence sensors. These sensors are necessary to signal the robots or a dedicated machine that the pallet is in the right position and the workstation operation can be performed. The cycle time to transport the part from one workstation to the next is usually constant, making the workstations synchronous. In certain cases the production line developers integrate parts to buffer the pallets, making the production line asynchronous. The need for a buffer arises in cases where some workstations have variable cycle times; with a buffer the overall production line cycle time is not affected.

Another very common assembly line configuration is a rotational or rotary table (Fig. 11.5), usually actuated by electrical motors. The speed and repeatability of positioning are high. The rotary table is often called a dial table or an indexing machine. The advantage of the rotary configuration is that requires less floor space and is often cheaper than other production line configurations. The rotary table is always performing synchronous transfer of parts between workstations with a constant cycle time.

As with the previous example, this configuration can also be served by humans, robots or dedicated machines. The rotary table has a circular shape around which the pallets or part-holders are traveling and transporting parts, in turn, into each manual or automated workstation where production operations are performed. The rotary table can be split in several workstations (minimum 2), making the rotation angle of 90°. More common are rotary tables with more than 2 workstations, e.g. 4, 5, 6 workstations. The size of the rotary table is defined by the part size, equipment size and number of workstations of the rotary production line. Closed loop controlled turntables are also available.

Usually a combination of the above configurations is installed and is referred to as hybrid production line configuration. Several factors declare the overall configuration of the production line, such as:

- space needed for production line,
- cost of installation of the production line,
- production line cycle time.

11.3 Feeding Devices

The task of the feeding devices is to bring parts or assemblies to the robot or dedicated machine in such a way that the part pose is known. Reliable operation of the feeding devices is of utmost importance in the robot cells without robot vision. The position of a part must be accurate, as the robot end-effector always moves along the same trajectory and the part is expected to be always in the same place.

The requirements for the robot feeding devices are much more strict than in manual assembly, unless the robot cell is equipped with a robot vision system. The robot feeding devices must not deform the parts, must operate reliably, position the parts accurately, work at sufficient speed, require minimal time of loading and contain sufficient number of parts.

The feeding device should not cause any damage to the parts handled, as damaged parts would afterwards be inserted by the robot into assemblies which cannot function properly. The cost of such damaged assemblies is higher than the cost of a more reliable feeding device. The feeding device must reliably handle all the parts whose dimensions are within tolerance limits. It must also be fast enough to meet the requirements of the whole production line cycle time and should never slow down its operation. Further, the feeding device should require as little time as possible for loading of the parts. It is more desirable to fill a large amount of parts into the feeding device at once than inserting them manually one by one. The feeding devices should contain as large number of parts as possible. This way the number of loadings required per day is reduced.

The simplest feeding devices are pallets and fixtures; an every-day example is the carton or plastic pallets used for eggs. The pallets store the parts, while determining their position and sometimes also orientation. In an ideal situation the same pallet is used for shipping the parts from the vendor and for later use in the consumer's robot cell. The pallets are either loaded automatically by a machine or manually. Fragile parts, flexible objects or parts with odd shapes must be loaded manually. Loading of

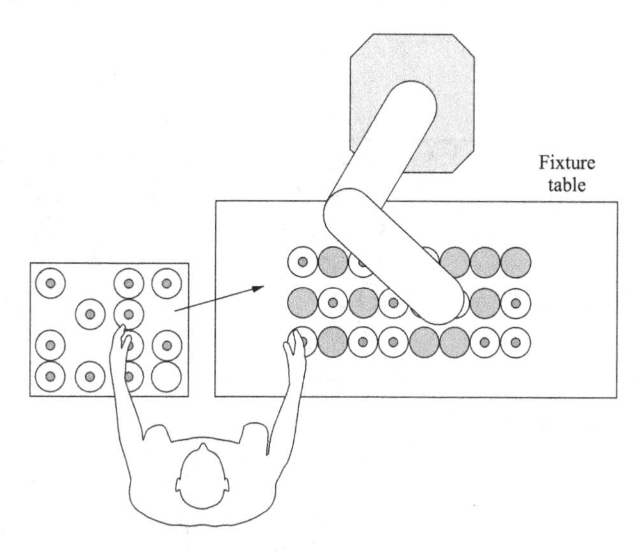

Fig. 11.6 Simultaneous loading of a fixture table

the pallets represents the weakest point of palletizing. Another disadvantage of pallets is their rather large surface, taking up considerable area in the robot workspace.

The simplest way to bring parts into the robot cell is represented by a fixture table. The human operator takes a part from a container, where the parts are unsorted, and places it onto the fixture table inside the robot workspace (Fig. 11.6). The fixture table must contain special grooves which assure reliable positioning of a part into the robot workspace. Such a fixture table is often used in welding where the component parts must be also clamped onto the table before the robot welding takes place. The time required for robot welding is considerably longer than loading and unloading which can justify the use of a fixture table.

The pallets can be loaded in advance in some other place and afterwards brought into the robot cell (Fig. 11.7). This avoids a long waiting period for the robot while the human operator is loading the pallets. The human worker must only bring the pallet into the robot workspace and position it properly using special pins in the working table. It is important that the pallet contains a sufficient number of the parts to allow continuous robot operation. Exchanging the pallets in the robot workspace represents a safety problem as the operator must switch off the robot or the robot cell must be equipped with other safety solutions (e.g. rotary table or collaborative robot).

A larger number of pallets can be placed on a rotary table (Fig. 11.8). The rotary table enables loading of the pallets on one side, while the robot activities take place on the other side of the turntable. This way robot cell inactivity is considerably reduced and the human operator is protected against the movements of the robot.

There are generally three types of pallets used: vacuum formed or, injection molded plastic and metal pallets. Since the cost of vacuum formed pallets is low,

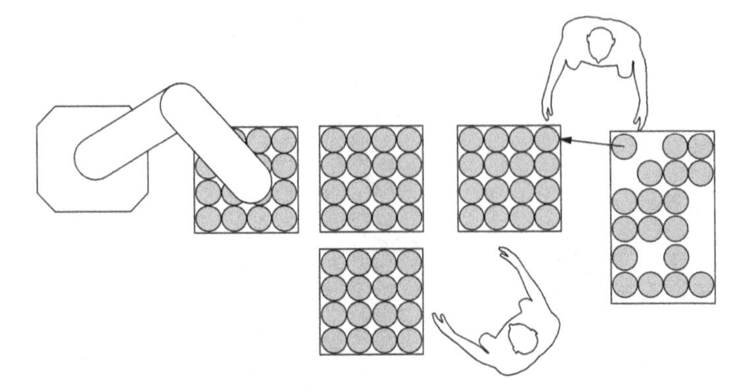

Fig. 11.7 Loading of the pallets in advance

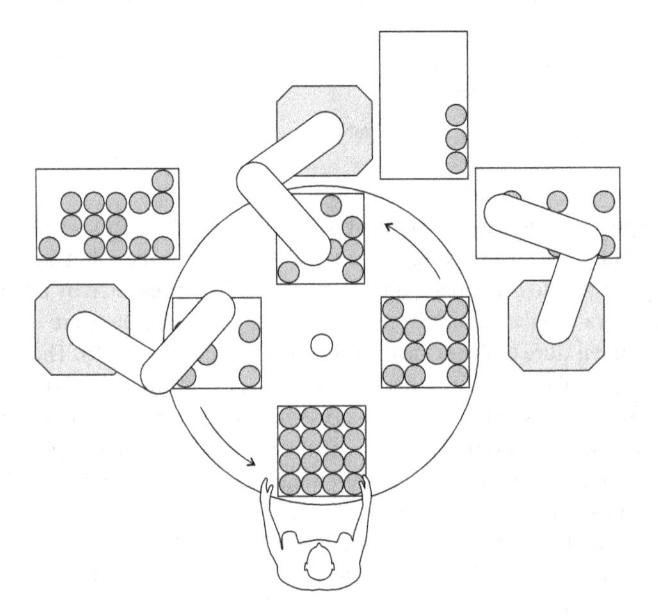

Fig. 11.8 Rotary table with pallet stages

they are used both for packaging and shipping of the parts and for use in the robot cell. Reference holes must be built into the all pallets to match pins in the worktable to enable simple and fast positioning. As the vacuum formed pallets are inexpensive, it is not difficult to understand that they are not the most accurate, reliable, or durable. They are made of a thin sheet of plastic material which is heated and vacuum formed over a mold. The inaccuracy of the pallet is the consequence of its low rigidity. Injection molded plastic pallets are used when more accurate and more durable pallets are required. The production of the mold is rather expensive, while the cost of production of a single pallet is not high. We must keep in mind that most vacuum and molded plastic pallets are flammable. Metal pallets are the only ones

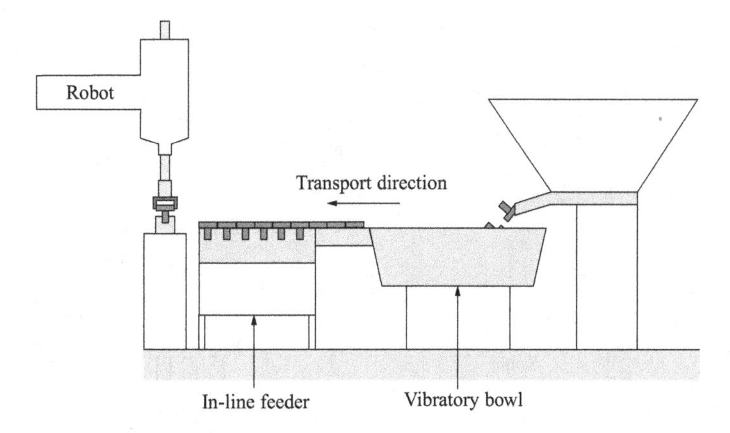

Fig. 11.9 Vibratory bowl feeder

which are non-flammable. They are produced by various machining approaches. The metal pallets are the most reliable and durable, while their cost is higher than that of plastic type. They are therefore only used inside the robot assembly process.

Part feeders represent another interesting family of feeding devices which are used not only for storing parts, but also for positioning and even orienting them into the pose appropriate for robot grasping. The most common are vibratory bowl feeders (Fig. 11.9). Here, the parts are disorderly loaded into the bowl. The vibration of the bowl and the in-line feeder is produced by an electromagnet, and the proper vibration is obtained by attaching the vibratory feeders to a large mass, usually a thick steel table. The vibrations cause the parts to travel out of the bowl. Specially formed spiral shaped fences force them into the required orientation. The same bowl feeder can be used for different parts, however not at the same time. Another benefit is that the bowl can hold a large number of parts while occupying only a small area in the robot workspace. Bowl feeders are not appropriate for parts such as soft rubber objects or springs. Another disadvantage is possible damage caused by the parts becoming jammed in the bowl. The noise of vibratory feeders can also be disturbing.

A simple magazine feeder consists of a tube storing the parts and the sliding plate, pneumatically or electrically actuated, which takes the parts one by one out of the magazine (Fig. 11.10). The magazine is loaded manually, so that the orientation of the parts is known. Gravity pushes the parts into the sliding plate. The mechanism of the sliding plate must be designed in such a way that it prevents jamming of the parts, while only a single part is fed out from the feeder at a time. The sliding plate must block all the parts except the bottom one.

Magazine feeders are excellent solutions for handling integrated circuits (Fig. 11.11). Integrated circuits are already shipped in tubes which can be used for feeding purposes. The magazine feeder for integrated circuits usually consists of

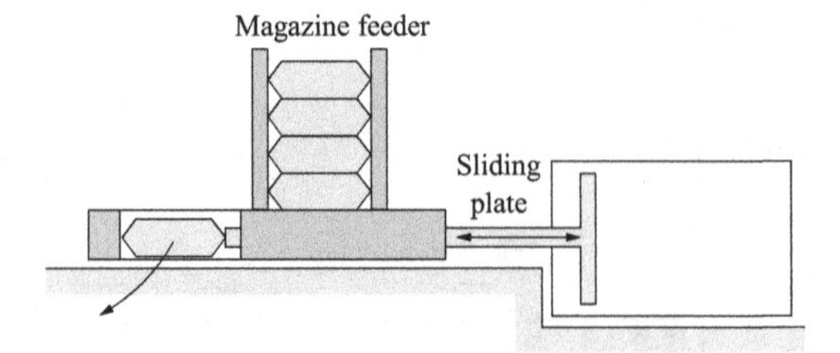

Fig. 11.10 Magazine feeder

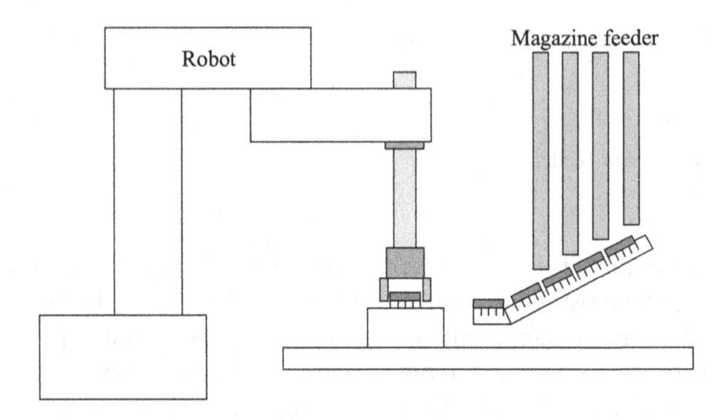

Fig. 11.11 Integrated circuit magazine feeder

several tubes. The tubes are aligned along a vibratory in-line feeder. The main disadvantage of magazine feeders is manual loading. They are also inappropriate for handling large objects.

11.4 Conveyors

Conveyors are used for transport of parts, assemblies or pallets between the robot cells. The simplest conveyor makes use of a plastic or metal chain which pushes the pallets along a metal guide (Fig. 11.12). An electrical motor drives the chain with constant velocity. The driving force is represented by the friction between the chain and the pallet. The pallet is stopped by special pins actuated by pneumatic cylinders. The chain continues to slide against the bottom of the pallet. When another pallet arrives, it is stopped by the first one. This way a queue of pallets is obtained in front of the robot cell.

The turn of a conveyor is made by bending the metal guide. The advantages of the sliding chain conveyor are low cost and simplicity in handling the pallets and performing the turns. The disadvantage is that perpendicular intersections cannot be made. Also, the turns must be made in wide arcs, which takes considerable floor

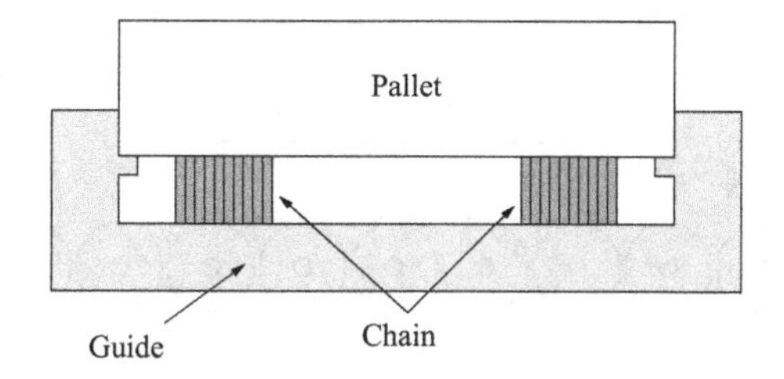

Fig. 11.12 Sliding chain conveyor (end view)

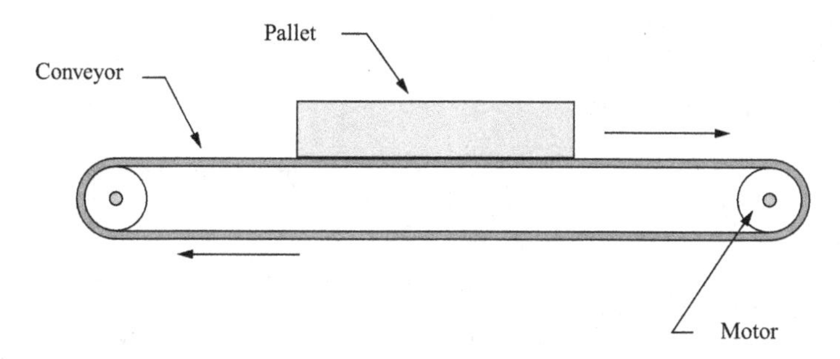

Fig. 11.13 Belt conveyor

space in the production facility. The sliding chain conveyor is best suited when used as a single loop feeding system.

With the belt-driven conveyor, the upper part of the belt is driving pallets or other objects or material (Fig. 11.13). A turn or intersection is made with the help of a special device enabling lifting, transfer and rotation of pallets.

A conveyor can also consist of rollers which are actuated by a common driving shaft (Fig. 11.14). The driving shaft transmits torque through a drive belt to the roller shaft. The advantage of the conveyor with rollers is in low collision forces occurring between the pallets or objects handled by the conveyor. They are the consequence of low friction between the rollers and the pallets. The turns are made by the use of lift and transfer devices. The disadvantages of the conveyors with rollers are high cost and low accelerations.

11.5 Robot Grippers and Tools

In the same way as robot manipulators are copies of the human arm, robot grippers imitate the human hand. In most cases robot grippers are considerably simpler than the human hand, encompassing wrist and fingers, altogether 22 degrees of freedom. Industrial robot grippers differ to a large extent, so it is not difficult to understand that their cost range from almost negligible to higher than the cost of a robot manip-

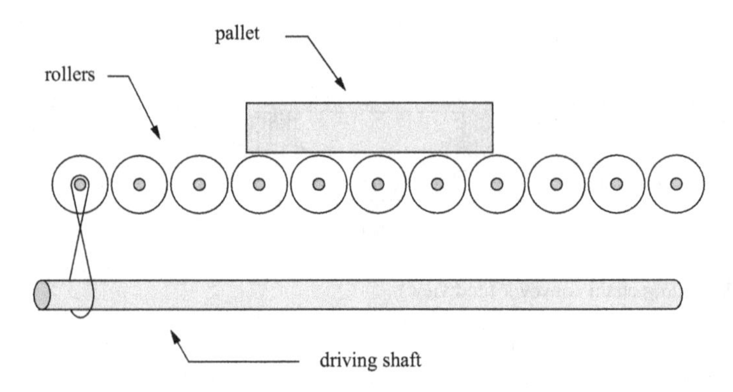

Fig. 11.14 Conveyor with rollers

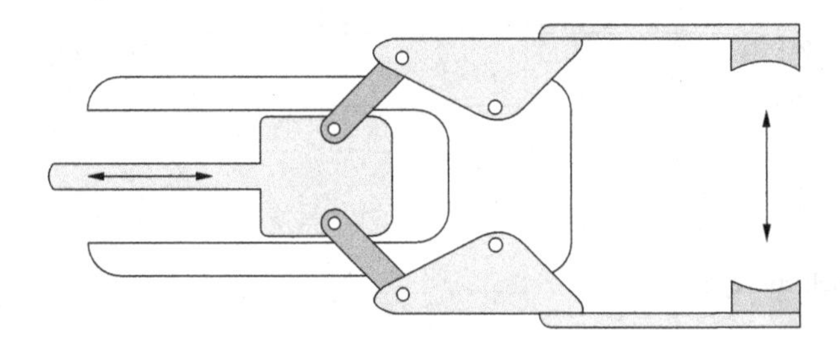

Fig. 11.15 Robot gripper with two fingers

ulator. Although many various robot grippers are commercially available, it is often necessary to develop a special gripper to meet the requirements of a specific robotic task.

The most characteristic robot grippers are those with fingers. They can be divided into grippers with two fingers (Fig. 11.15) and multi-fingered grippers. Most multi-fingered grippers have three fingers (Fig. 11.16), to achieve a better grasping. In industrial applications we usually encounter grippers with two fingers. The simplest two-finger grippers are only controlled between the two states, open and closed. Two-finger grippers, where the distance or force between the fingers can be controlled, are also available. Multi-fingered grippers usually have three fingers, each having three segments. Such a gripper has 9 degrees of freedom which is more than robot manipulator. The cost of such grippers is high. In multi-fingered grippers the motors are often not placed into the finger joints, as the fingers can become heavy or not strong enough. Instead, the motors are all placed into the gripper palm, while tendons connect them with pulleys in the finger joints. Apart from grippers with fingers, in industrial robotics there are also vacuum, magnetic, perforation and adhesive grippers. Different end-effector tools, used in spray painting, finishing or welding, are not considered robot grippers.

Two-fingered grippers are used for grasping the parts in a robotic assembly process. An example of such a gripper is shown in Fig. 11.15. Different end-points can

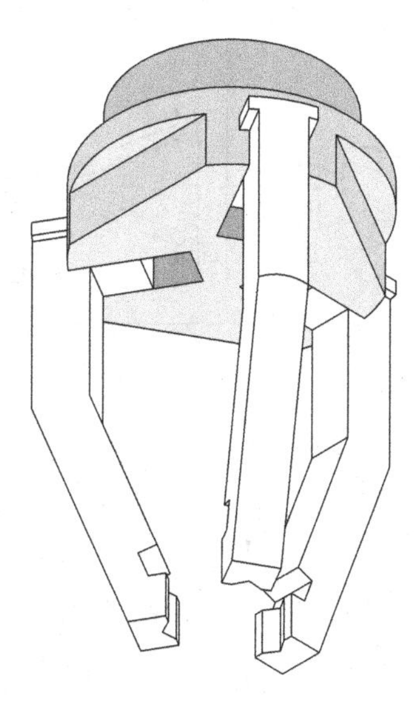

Fig. 11.16 Robot gripper with three fingers

be attached to the fingers to adapt the robot grasp to the shape and surface of the part or assembly to be grasped. With two-fingered robotic grippers pneumatic, hydraulic or electrical motors are used. Hydraulic actuation enables higher grasping forces and thus handling of heavier objects. Different structures of two-fingered grippers are presented in Fig. 11.17. Simple kinematic presentations enable the choice of an appropriate gripper for the selected task. The gripper on the right side of Fig. 11.17 enables parallel finger grasping.

In industrial processes, robot manipulators are often used for machine loading. In such cases the robot is more efficient when using a twofold gripper. The robot can simultaneously bring an unfinished part into the machine while taking a finished part out of it. A twofold gripper is shown in Fig. 11.18.

Specific grippers are used for grasping hot objects. Here, the actuators are placed far from the fingers. When handling hot objects air cooling is applied, while often the gripper is immersed into water as part of the manipulation cycle. Of utmost importance is also the choice of appropriate material for the fingers.

When grasping lightweight and fragile objects, grippers with spring fingers can be used. This way the maximal grasping force is constrained, while at the same time it enables a simple way of opening and closing of the fingers. An example of a simple gripper with two spring fingers is shown in Fig. 11.19.

The shape of the object requires careful design of a two-fingered robot gripper. A reliable grasp can be achieved either by form or force closure of the two fingers. Also possible is the combination of the two grasp modes (Fig. 11.20).

When executing a two-fingered robot grasp, the position of the fingers with respect to the object is also important. The grasping force can be applied only on the external

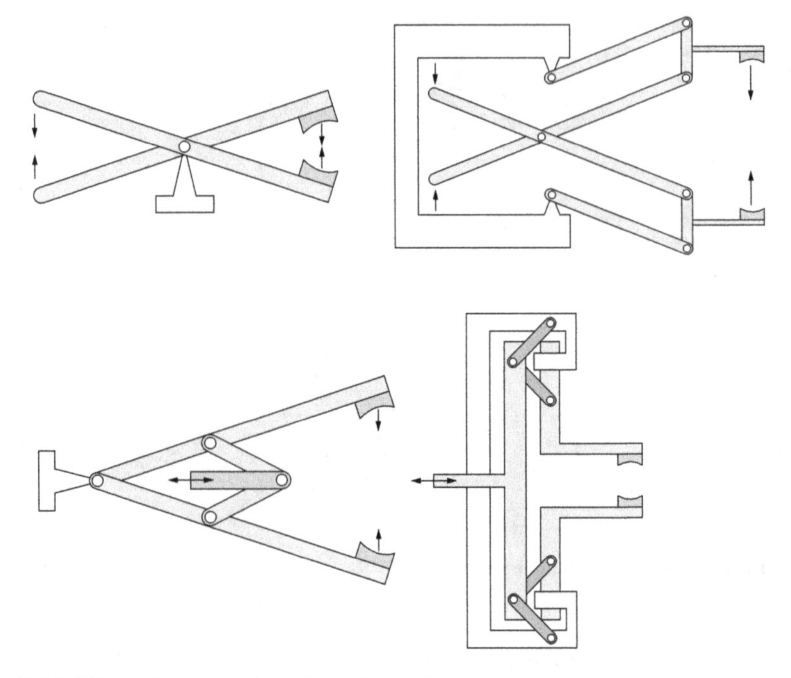

Fig. 11.17 Kinematic presentations of two-fingered grippers

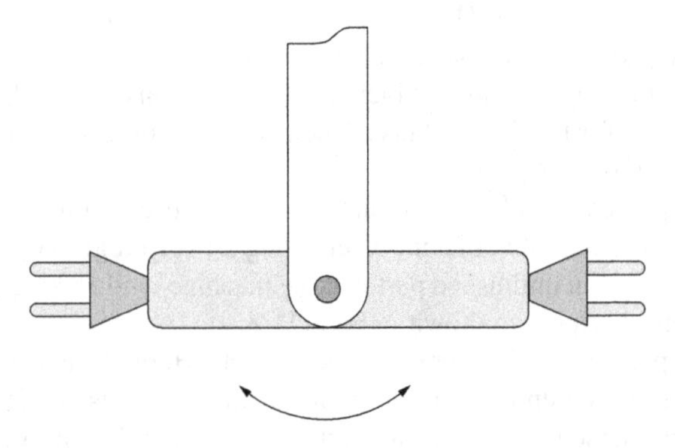

Fig. 11.18 Twofold robot gripper

surfaces or only on the internal surfaces of a work-piece. An intermediate grip is also possible where the object is grasped on internal and external surfaces (Fig. 11.21).

Among the robot grippers without fingers, vacuum grippers are by far the most frequently used. Vacuum grippers or grippers with negative pressure are successfully applied in cases, where the surface of the grasped object is flat or evenly curved, smooth, dry and relatively clean. The advantages of these grippers are reliability, low cost and small weight. Suction heads of various shapes are commercially available. Often several suction heads are used together, being put into a pattern that suits the

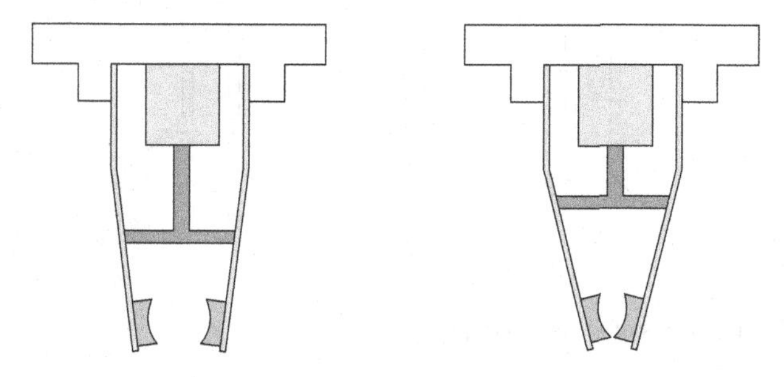

Fig. 11.19 Gripper with spring fingers

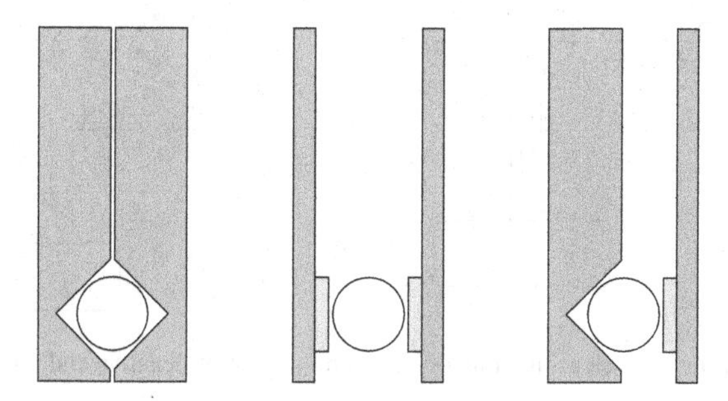

Fig. 11.20 Form closure, force closure and combined grasp

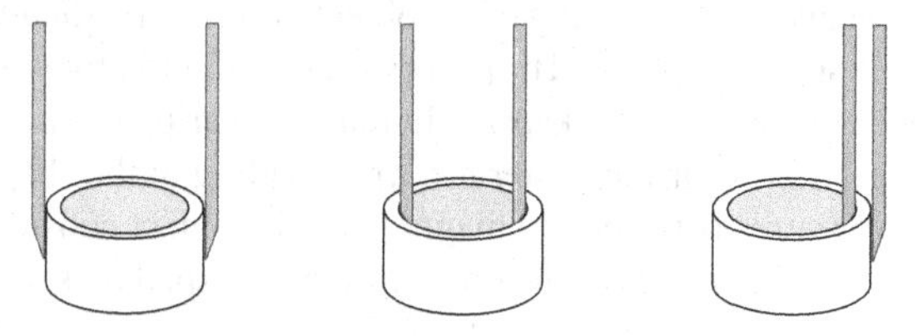

Fig. 11.21 External, internal, and intermediate grip

shape of the object to be grasped. Figure 11.22 shows the shape of two frequently used suction heads. The head on the left is appropriate in cases when the surface is not completely smooth. The soft material of the head adapts to the shape of the object. The small nipples on the head presented on the right side of Fig. 11.22 prevent damage to surface of the object. Vacuum is produced either with Venturi or vacuum pumps. The Venturi pump needs more power and produces only 70% vacuum. However, it is often used in industrial processes because of its simplicity and low cost. Vacuum pumps provide 90% vacuum and produce considerably less noise. In all grippers, fast grasping and releasing of the objects is required. Releasing very lightweight and sticky objects can be critical with vacuum grippers. In this case we release the objects with the help of positive pressure as demonstrated in Fig. 11.23.

Magnetic grippers are another example of grippers without fingers: these use either permanent magnets or electromagnets. The electromagnets are used to a larger extent.

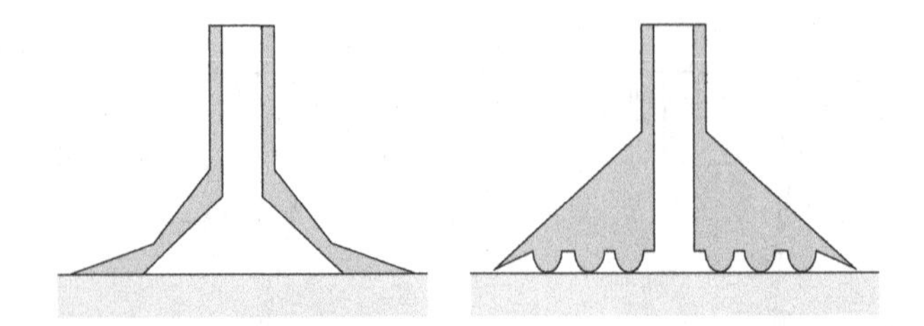

Fig. 11.22 Suction heads of vacuum grippers

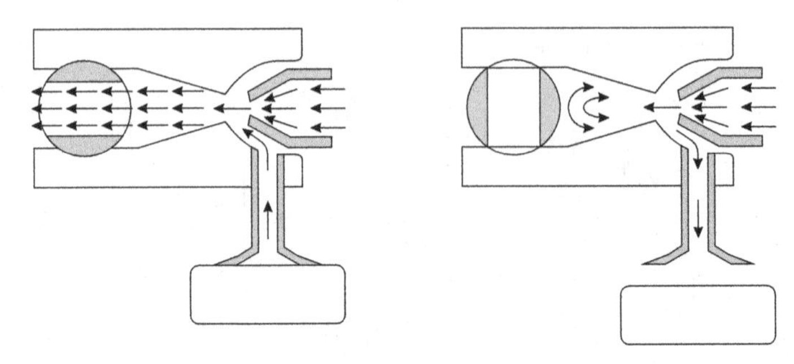

Fig. 11.23 Grasping and releasing of an object with the help of negative and positive pressure

With permanent magnets the releasing of the object presents a difficulty. The problem is solved by using a specially planned trajectory of the end-effector where the object is retained by a fence in the robot workspace. In magnetic grippers several magnets are used together, placed into various patterns corresponding to the shape of the object. Already small air fissures between the magnet and the object considerably decrease the magnetic force. The surfaces of the objects being grasped must be therefore even and clean.

Perforation grippers are considered as special robot grippers. Here the objects are simply pierced by the gripper. Usually these are used for handling material such as textile or foam rubber. Such grippers can be used only in cases when perforation does not cause damage to the object. Sheets of textile can be grasped by large brushes made of stiff nylon hairs or simply of Velcro straps.

Adhesive grippers can be used when grasping very lightweight parts. Release of the parts must be solved by special robot end-point trajectories where the part collides with the fence in the robot workspace and is thus removed from the adhesive gripper. Sufficient adhesive force is provided using adhesive tape which must move during the operation.

Besides grippers the robot can have other tools attached to its end. The shape and the function of the tool depends on the task of the robot cell. The most frequent operation that robots perform is welding. For welding purposes several different approaches can be used. Among them the most frequent tool attached to the robot

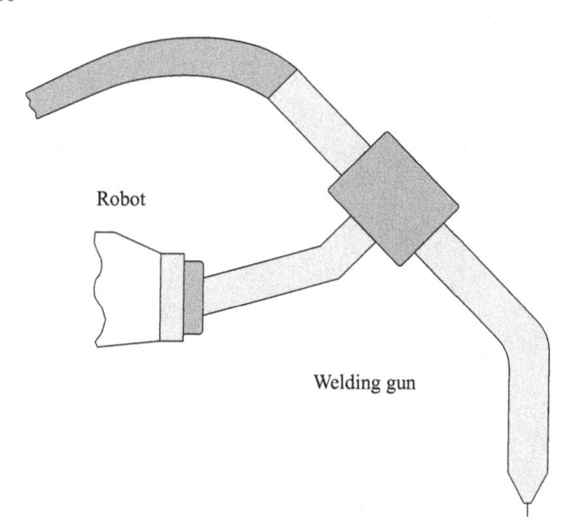

Fig. 11.24 Robot with welding gun attached to its end

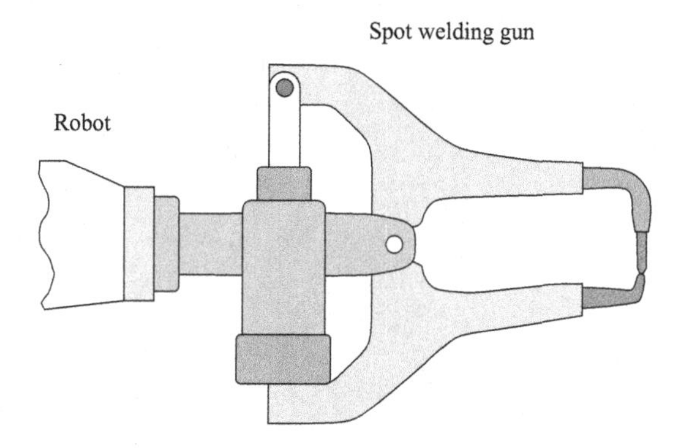

Fig. 11.25 Robot with spot welding gun attached to its end

end is an arc welding gun or torch (Fig. 11.24) to transmit welding current from a cable to the electrode. The task is performed in many different areas of manufacturing. Besides arc welding also spot welding guns (Fig. 11.25) can frequently be found in manufacturing processes, mainly in the automotive industry.

Chapter 12
Collaborative Robots

In 1942 Isaac Asimov published the science fiction novel "I, Robot", where the three laws of robotics were introduced. First rule stated that "A robot may not injure a human being or, through inaction, allow a human being to come to harm".

Until now, industrial robots have always been fast and robust devices that work on specific tasks designed for them. To stay in accordance with the aforementioned rule they were performing behind fixed and interlocked guards and sensitive protective equipment to prevent human intrusion into their workspace. With the introduction of collaborative robots the cages are omitted as those robots are designed to work with humans. They are built with different safety features to prevent collisions, but if a collision occurs, the mechanism will move in the opposite direction or stop completely to avoid causing injury.

The technical specification ISO/TS 15066:2016: Robots and robotic devices—Collaborative robots supplements the requirements and guidance on collaborative industrial robot operation provided in ISO 10218-1:2011 and ISO 10218-2:2011 (ANSI/RIA R15.06:2012). It specifies safety requirements for collaborative industrial robot systems and the work environment. Specifically, ISO/TS 15066:2016 provides comprehensive guidance for risk assessment in collaborative robot applications.

12.1 Collaborative Industrial Robot System

A collaborative robot is a robot that can be used in a collaborative operation, where a purposely designed robot system and a human operator work in direct cooperation within a defined workspace. The term robot defines robot arm and robot control and does not include the robot end-effector or part. With the term robot system we describe robot, end-effector, and workpiece.

For the collaborative robot system we can define different workspaces (Fig. 12.1):

M. Mihelj et al., *Robotics*, https://doi.org/10.1007/978-3-319-72911-4_12

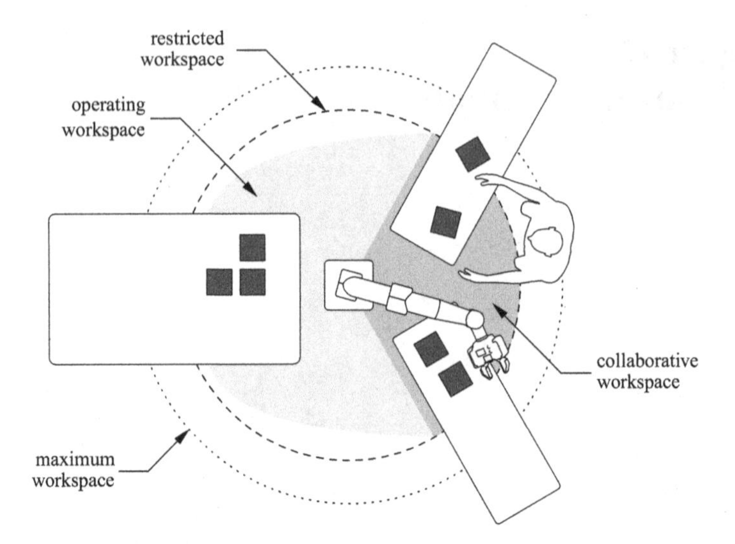

Fig. 12.1 Maximum workspace (limited by dotted line), restricted workspace (limited by dashed line), operating workspace (grey areas), and collaborative workspace (dark grey area)

- maximum workspace: space which can be swept by the moving segments of the robot as defined by the manufacturer plus the space which can be swept by the end-effector and the workpiece;
- restricted workspace: portion of the maximum space restricted by limiting devices that establish limits which will not be exceeded;
- operating workspace: portion of the restricted space that is actually used while performing all motions commanded by the task program;
- collaborative workspace: portion of the operating space where the robot system and a human can perform tasks concurrently during production operation.

The collaborative workspace must be designed in a way that the operator can perform all intended tasks. The location of machinery and equipment should not introduce any additionally safety hazards. In the collaborative workspace strict limitations about the speed, space limits, and torque sensing are applied to guarantee operator safety. Outside the collaborative workspace the robot can act as a traditional industrial robot without any particular limitations excluding those that are task-related.

The term operator includes all personnel that are in contact with the robot system, not only production operators. It includes maintenance, troubleshooting, setup, cleaning, and production personnel.

The operational characteristics of collaborative robot systems are significantly different from those of traditional industrial robot system presented in ISO 10218-1:2011 and ISO 10218-2:2011. In collaborative robot operations, operators can work in direct proximity to the robot system while the system is active, and physical contact between an operator and the robot system can occur within the collaborative workspace. As such, adequate protective measures must be introduced to collaborative robot systems to ensure the operator's safety at all times during collaborative robot operation.

12.2 Collaborative Robot

The design of collaborative robots is moving away from heavy, stiff, and rigid industrial robots towards lightweight devices with an active and/or passive compliance. The use of lightweight high-strength metals or composite materials for robot links contributes to small moving inertia which further affects the power consumption of the motors. Serial manipulators can be equipped with high power/torque motors with high transmission ratio gears in each joint or have motors positioned at the base while the power is transferred via tendons. If the transmission ratio is small the system is inherently back-drivable.

Use of intrinsically flexible actuators enables the design of biologically inspired robots, as the actuators mimic the performances of human/animal muscles. The actuators can have fixed mechanical impendence controlled via active control, such as series elastic actuator (SEA), or the impedance can be adjusted by changing parameters of a mechanical joint, as in variable stiffness actuator (VSA). SEA is a combination of motor, gearbox, and a spring, where the twist of the spring is measured to control the force output, while that measurement of the twist of the spring is used as a force sensor. VSA can be used to make the robot safer in the case of collision as the joint stiffness and impact inertia are reduced. Conceptual designs of SEA and VSA are presented in Fig. 12.2.

Collaborative robots also have special geometries that minimize the contact energy transfer by maximizing the impact area. Robots have round shapes and integrated features that reduce the risk of pinch points and the severity of an impact. Main features of the collaborative robot are presented in Fig. 12.3

To ensure a high level of safety, the robot system must include different sensors for monitoring the state of the robot and its workspace as presented in Chap. 7. Robots can be equipped with joint torque sensors, force/torque sensors at the end-effector, and different tactile sensors used as a soft skin or a hard shell for the robot. All these sensors enable the robot to detect contact with the environment (operator) or avoid collision by anticipating it and responding accordingly. Some robots use

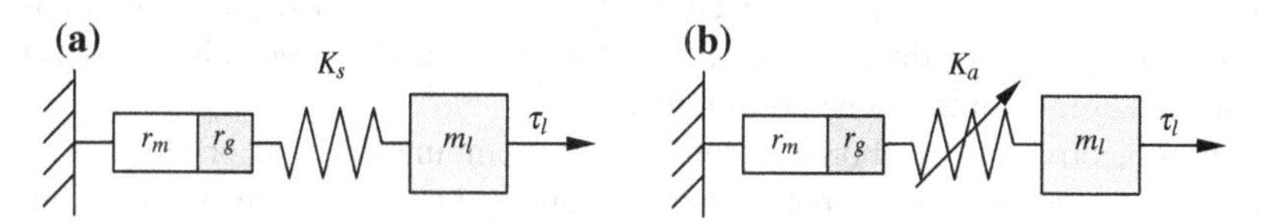

Fig. 12.2 **a** Series elastic actuator (SEA), **b** variable stiffness actuator (VSA); r_m and r_g represent motor and gearbox, K_s compliant element with fixed stiffness, K_a adjustable compliant element, m_l moving link's mass, and τ_l joint torque resulting in link movement

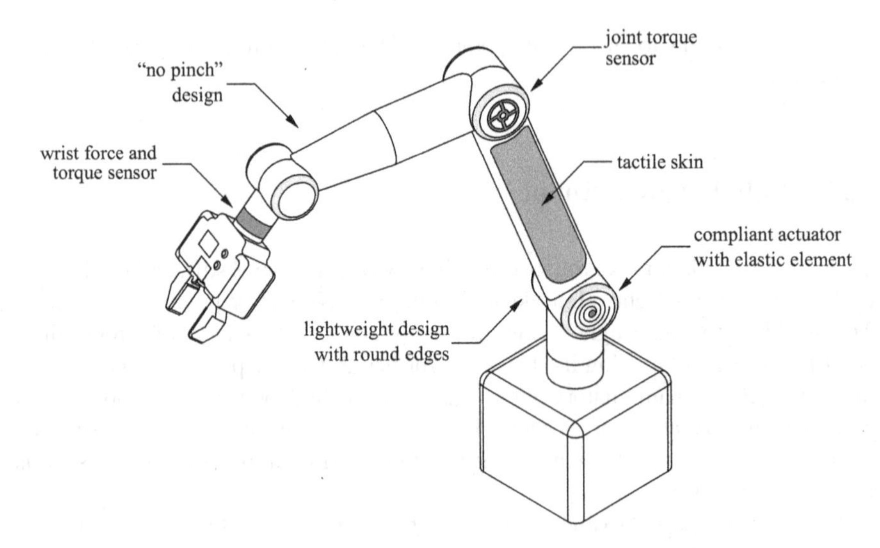

Fig. 12.3 Design features of a collaborative robot

redundant encoders in every joint to substitute for expensive joint torques; force can be derived from the known motor current and joint position. Robot systems can include other safety rated sensors, e.g., safety cameras, laser scanners, laser curtains, safety mats and other electro-sensitive protective equipment, to detect the presence of the operator in the robot surroundings. This information can be then used for a proper robot response to prevent clamping, compression, or crushing of the operator.

The incorporated sensors can be used for safe control of the robot. The main paradigm is how to handle physical contact between the mechanism and the surroundings. One of the most popular control schemes is impedance control, that is based on the dynamic robot model (5.56). The dynamic model is used to assess the necessary joint torques for proper robot movement. If the measured joint torques deviate from the assessed one, then the difference is detected as a collision. When a collision has been detected, the proper response strategy should be activated to prevent potential danger to the operator. The robot can ignore the contact and follow the reference trajectory, or the robot can be stopped. Other possibilities include switching from position control to zero-gravity torque control (very high compliancy of the robot), switch to torque control with the use of signals from joint torques to minimize link and motor inertia (even "lighter" robot), or to use external measured torques and switch to admittance control, where robot and collided object act as two magnets facing with the same poles together.

The objective of collaborative robots is to combine the best of robots and of human operator: the robot's precision, power, and endurance coupled with the human operator's excellent capability for solving imprecise problems. As the robot and the operators are collaborating in the same workspace, contact between robots and humans is allowed. If an incidental contact does occur, then that contact should not result in pain or injury. As such, collaborative robots can be used alongside operators and enhance the productivity of the workers. Robots are lightweight and have a small footprint so can be easily moved around workshop, thus increasing their versatility. Programming of collaborative robots is simple, mostly done by hand guiding, so the use of the robot is very flexible; the robot can be operational at a new workstation in a very short time.

12.3 Collaborative Operation

Collaborative operation is not defined with the use of the robot alone but is conditioned by the task, what the robot system is doing, and the space in which the task is being performed. Four main techniques (one or combination of more) can be included into collaborative operation:

- safety-rated monitored stop;
- hand guiding;
- speed and separation monitoring;
- power and force limiting.

With all four techniques the robot performs in automatic mode. The main details of all four methods are presented in Table 12.1. More detailed descriptions are available further below.

Table 12.1 Types of collaborative operations

	Speed	Torques	Operator controls	Technique
Safety-rated monitored stop	Zero while operator is in collaborative workspace	Gravity and load compensation only	None while operator is in collaborative workspace	No motion in the presence of the operator
Hand guiding	Safety-rated monitored speed	As by direct operator input	Emergency stop, enabling device, motion input	Motion only by direct operator input
Speed and separation monitoring	Safety-rated monitored speed	As required to maintain min. separation distance and to execute the application	None while operator is in collaborative workspace	Prevented contact between the robot system and the operator
Power and force limiting	Max. determined speed to limit impact forces	Max. determined torque to limit static forces	As required by application	Robot cannot impart excessive force (by design or control)

12.3.1 Safety-Rated Monitored Stop

In this method the robot system must be equipped with safety-rated devices which detect the presence of the operator inside the collaborative workspace (e.g., light curtains or laser scanners). The operator is permitted to interact with the robot system in the collaborative workspace only when the robot's safety-rated monitored stop function is active and the robot motion is stopped before the operator enters the shared workspace. During collaborative task the robot is in standstill with the motors powered. Robot system motion can resume only when the operator has exited the collaborative workspace. If there is no operator in the collaborative workspace, the robot may operate as classical industrial robot, e.g., non-collaboratively.

The operations of the safety-rated monitored stop are presented in Table 12.2. When the operator is outside the collaborative workspace the robot can perform without any limitations. But in the case that the robot is present in the workspace at the same time as the operator, the robot's safety-rated monitored stop should be active. Otherwise the robot must engage category 0 protective stop (uncontrolled stop of the robot by immediately removing power to the actuators) in case of fault (IEC 60204-1).

This method can be applied to applications of manual loading or unloading of end-effector, work-in-progress inspections, and applications where only one moves in collaborative workspace, (e.g., robot or operator). Safety-rated monitored stops can also be integrated with other collaborative techniques.

12.3.2 Hand Guiding

For hand guiding the robot must be equipped with a special guiding device located at or near the robot end-effector that serves for transmitting motion commands to the robot system. The device must incorporate an emergency stop and an enabling device unless the robot system meets inherently safe design measures or safety-limiting functions. The location of the guiding device should enable the operator to

Table 12.2 Robot actions for safety-rated monitored stop

		Operator's proximity to collaborative workspace	
		Outside	Inside
Robot's proximity to collaborative workspace	Outside	Continue	Continue
	Inside and moving	Continue	**Protective stop**
	Inside, safety-rated monitored stop	Continue	Continue

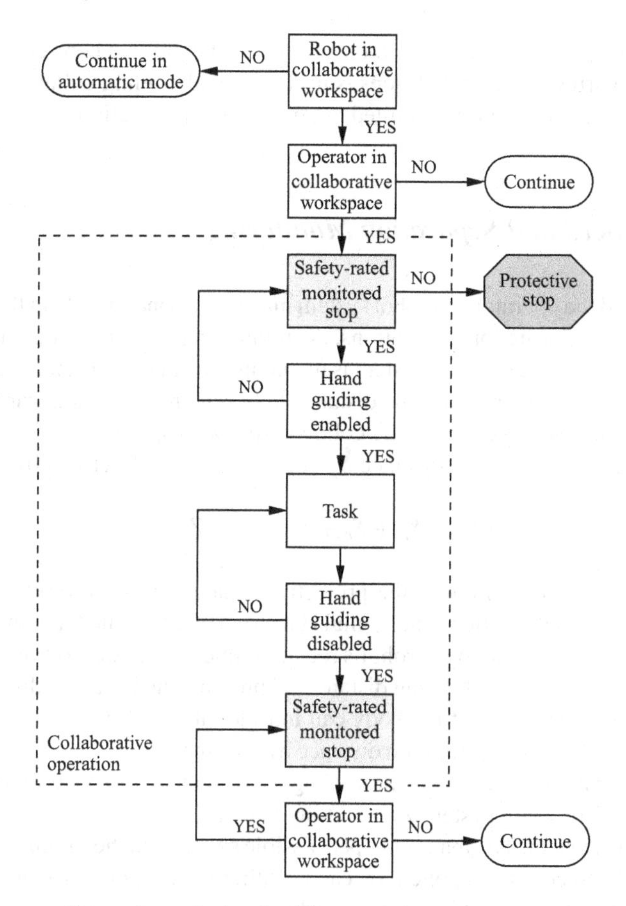

Fig. 12.4 The operating sequence for hand guiding

directly observe the robot motion and prevent any hazardous situations (e.g., operator is standing under heavy load). The control of the robot and end-effector should be intuitively understandable and controllable.

The robot system is ready for hand guiding when it enters the collaborative workspace and issues a safety-rated monitored stop. At this point the operator can enter the collaborative workspace and take control of the robot system with the hand guiding device. If the operator enters the collaborative workspace before the system is ready for hand guiding, a protective stop must be issued. After the safety-monitored stop is cleared the operator can perform the hand guiding task. When the operator releases the guiding device the safety-rated monitored stop is issued. Non-collaborative operation resumes when the operator leaves the collaborative workspace. The operating sequence for hand guiding is presented in Fig. 12.4.

This collaboration technique is suitable for implementation within applications where the robot system acts as a power amplifier, in highly variable applications, where robot system is used as a tool, and in applications where coordination of manual and partially automated steps is needed. Hand guiding collaboration can be successfully implemented into limited or small-batch productions.

12.3.3 Speed and Separation Monitoring

In this method the operator and robot system may move concurrently in the collaborative workspace. During joint operations, the minimum protective separation distance between the operator and robot system is maintained at all time. Protective separation distance is the shortest permissible distance between any moving hazardous part of the robot system and operator in the collaborative workspace.

The protective separation distance S_p at time t_0 can be described by (12.1):

$$S_p(t_0) = S_h + S_r + S_s + C + Z_d + Z_r \,, \tag{12.1}$$

where S_h is the contribution to the protective separation distance attributed to the operator's change in location. The formula takes into account the braking distance S_r, which is the distance due to the robot's reaction time, and S_s describing the distance due to the robot system's stopping distance. C presents the intrusion distance, which is the distance that a part of the body can intrude into the sensing field before it is detected. The protective separation distance S_p also includes the position uncertainty of the operator Z_d, resulting from the sensing measurement tolerance, and the position uncertainty of the robot system Z_r, resulting from the accuracy of the robot position measurement system. The maximum permissible speeds and the minimum protective separation distances in an application can be either variable or constant. The various contributions to the protective separation distance are illustrated in Fig. 12.5.

The robot must be equipped with a safety-rated monitored speed function and a safety-rated monitored stop. The robot system includes also additional safety-rated peripheral for human monitoring (e.g., safety-rated camera systems). The robot system can maintain minimum protective separation distance by speed reduction, which could be followed by safety-rated monitored stop, or execution of an alternate path which does not violate the protective separation distance, as presented in Fig. 12.6. If the actual separation distance between the robot system and the operator falls below the protective separation distance, the robot system should initiate a protective stop and initiate safety-related functions connected to the robot system (e.g., turn off any hazardous tools). When the operator moves away from the robot, the actual separation distance meets and exceeds the protective separation distance; at this point the robot can resume motion automatically.

Speed and separation monitoring is useful in applications where robot system's and operator's tasks run simultaneously.

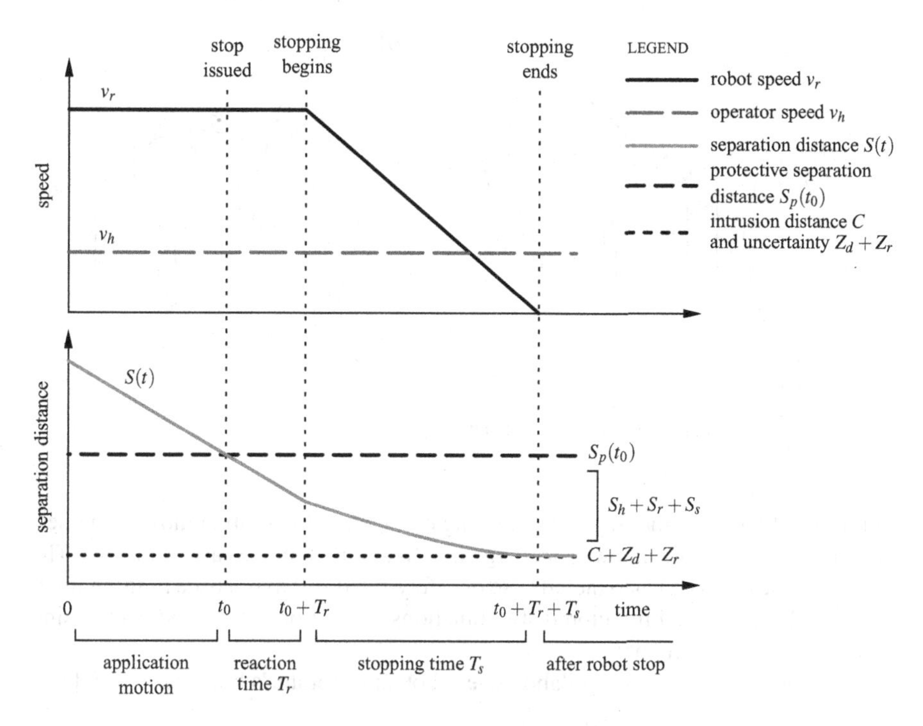

Fig. 12.5 Graphical representation of the contributions to the protective separation distance between an operator and a robot

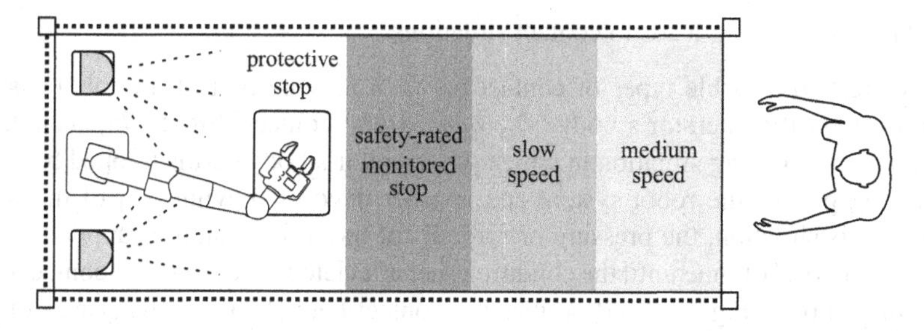

Fig. 12.6 Safety-rated levels for maintaining minimum protective separation distance

12.3.4 Power and Force Limiting

The method of power and force limiting allows physical contact between the robot system and the operator, that can occur either intentionally or unintentionally. The method demands that robots be specifically designed by means of low inertia, suitable geometry (rounded edges and corners, smooth and compliant surfaces), materials (padding, cushioning, deformable components), and control functions. The former includes active safety design methods, such as limiting forces and torques,

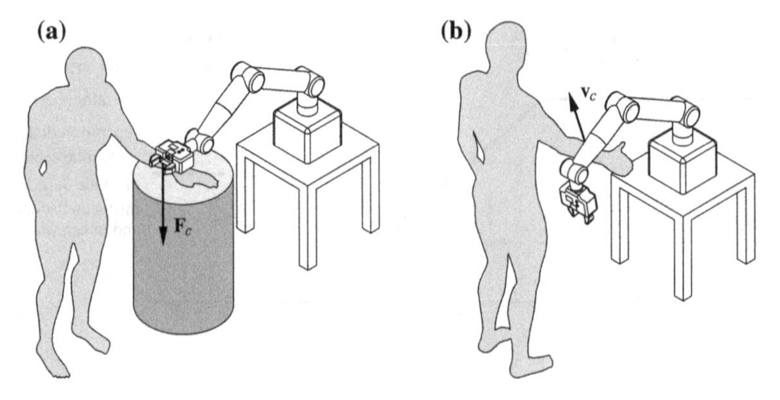

Fig. 12.7 **a** Quasi-static and **b** transient contact

limiting velocities of moving parts, limiting momentum by limiting moving masses, and limiting mechanical power or energy as a function of masses and velocities. The design of the robot can also include use of safety-rated soft axis, space limiting functions, and safety-rated monitored stop functions. Some robots also include sensing to anticipate or detect contact.

The contact between the collaborative robot and operator's body parts could be:

- intended as part of the application sequence;
- incidental due to not following the working procedure, but without technical failure;
- a failure mode that leads to contact situations.

There are two possible types of contact between moving part of the robot system and areas on the operator's body. The *quasi-static* contact (Fig. 12.7a) includes a clamping or crushing situation in which the operator's body part is trapped between a moving part of the robot system and another fixed or moving part of the work cell. In this situation, the pressure or force $\mathbf{F}_c$ of the robot system is applied for an extended period of time until the conditions are alleviated. The *transient* contact (i.e., dynamic impact, Fig. 12.7b) describes the contact between the moving part of the robot system and the operator's body part without clamping or trapping of that part. The actual contact is shorter than the aforementioned quasi-static contact ($<$50 ms), and depends on the inertia of the robot, the inertia of the operator's body part, and the relative speed $\mathbf{v}_c$ of the two.

The robot system must be adequately designed to reduce risk to an operator by not exceeding the applicable threshold limit values of force and pressure for quasi-static and transient contact. The limits can apply to forces, torques, velocities, momentum, mechanical power, joint ranges of motion, or space ranges. Threshold limit value for the relevant contact event on the exposed body region are determined for a worst-case scenario for both contact types.

The limit values presented in ISO/TS 15066:2016 are based on a conservative estimate and scientific research on pain sensations. Some informative values for

Table 12.3 Biomechanical limits for quasi-static contact

Body area	Maximum permissible pressure p_{QS}/N/cm^2	Maximum permissible force F_{QS}/N
Seventh neck muscle	210	150
Shoulder joint	160	210
Sternum	120	140
Abdomen	140	110
Pelvis	210	180
Humerus	220	150
Forearm	180	160
Palm	260	140
Forefinger pad	300	140
Forefinger end joint	280	140
Back of the hand	200	140
Thigh	250	220
Kneecap	220	220
Shin	220	130
Calf	210	130

maximum permissible pressure and maximum permissible force between the robot's part and operator's body region in quasi-static contact are presented in Table 12.3. Pressure and force values for transient contact (p_T, F_T) can be at least two times the values for quasi-static contact (p_{QS}, F_{QS}).

$$p_T = 2 \cdot p_{QS} \tag{12.2}$$

$$F_T = 2 \cdot F_{QS} . \tag{12.3}$$

Contact with face, skull, or forehead is not permissible and needs to be prevented.

For proper robot system reactions, both pressure and force limits must be taken into consideration, depending on the situation. In case of clamping of operator's body part (e.g., operator's hand), the resulting force can be well below the limit threshold so the pressure limit will be the limiting factor. On the other hand, if the contact is between two fairly large and soft areas (e.g., padded robot part and operator's abdomen), the resulting pressure will be below the limit threshold and the limiting factor will then be the force limit.

In case of contact, the robot system must react in a way that the effect of the identified contact remains below the identified threshold limit values, as presented in Fig. 12.8. In case of clamping or pinning a body part between a robot segment and some other object, the robot must limit the speed to comply with the protective limits. The robot should also have an integrated option for the operator to manually extricate the affected body area.

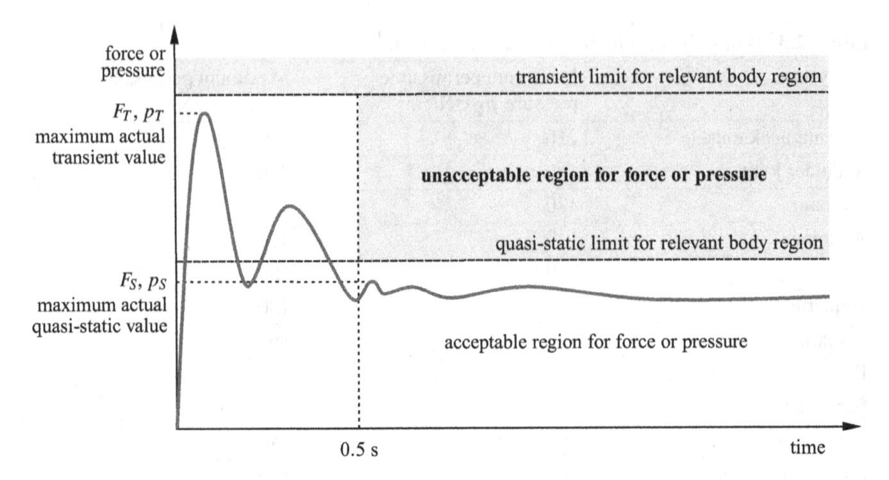

Fig. 12.8 Graphical representation of the acceptable and unacceptable forces or pressures in case of quasi-static or transient contact

The power and force limiting method can be used in collaborative application where the presence of the operator is frequently needed, in time-dependant operations (where delay due to safety-rated stops is unwanted but physical contact between the robot system and the operator can occur), and applications with small parts and high variability of assembly.

12.4 Collaborative Robot Grippers

The design and control of a collaborative robot enables the robot to be safe while working together with the operator. But the robot itself is just a part of the robot system. Grippers represent an important part of the robot system as they are used for object manipulation in the direct vicinity of the operator. As such, grippers must attain high level of safety.

The grippers are usually rigidly attached to the already-safe robot with built-in speed and force limitations. The shape and materials of the gripper must coincide with the safety design preventing exceeded pressure limits on the contact area of the operator's body. In addition, the grippers at the tip of the robot should create as little inertia as possible to minimally interfere with robot's safety features.

The design of the grippers should prevent the operator from getting their fingers stuck in the gripper or in the connecting cables. The grippers must have implemented a safe mode under an emergency stop, which function depends on the application. If there is a gripped part, the operator usually wants the part to stay safely gripped. When teaching and closing the gripper, the operator wants the gripper to stop applying the force.

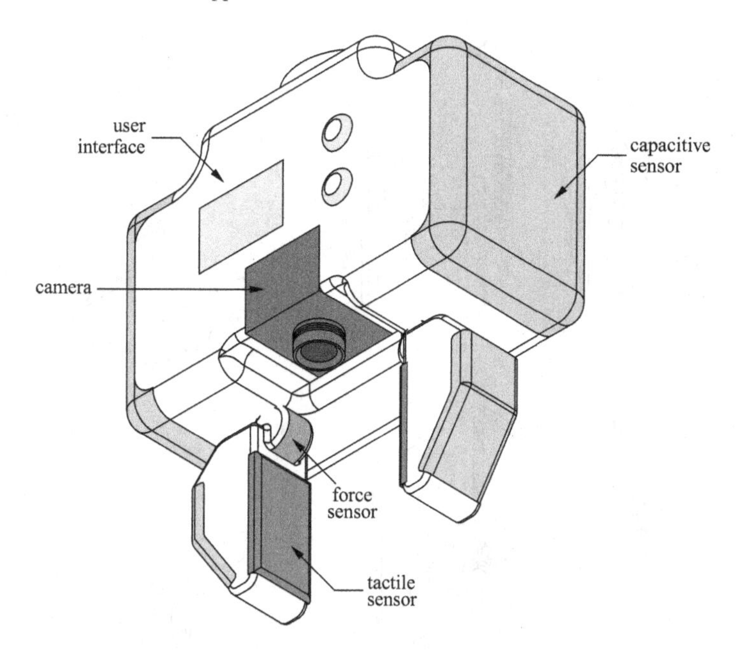

Fig. 12.9 Conceptual design of a gripper for collaborative gripping

When the gripper is interacting with the part, the operator wants a good solid grip. The grip also has to be secure under an emergency stop or power loss as a dropped part could represent a danger for an operator, robot, or environment. If the robot is moving fast, the dropped part could become a projectile.

Grippers can be equipped with different sensors to increase the operator's safety (Fig. 12.9). Capacitive sensors are used for early operator detection and thus prevention of unwanted contacts. Camera systems can detect the robot's surroundings and aids in object search. Tactile sensors are used to differentiate between workpiece and operator. To set adequate gripping force, different force sensors can be integrated. The gripper design can also include different user interfaces, such as LCD screen, signal lights, and control buttons.

Grippers used in the collaborative robot systems should be easy to install and program. The future design of the grippers is tending away from user programming towards grippers that will be capable of automatic adaptation depending of the parts and applications.

12.5 Applications of Collaborative Robotic System

The document ISO 10218-2:2011 provides the division of collaborative applications into five categories presented in Fig. 12.10.

Hand-over window application (see Fig. 12.10a) covers loading/unloading, testing, benching, cleaning, and service tasks. The robot is positioned behind fixed or

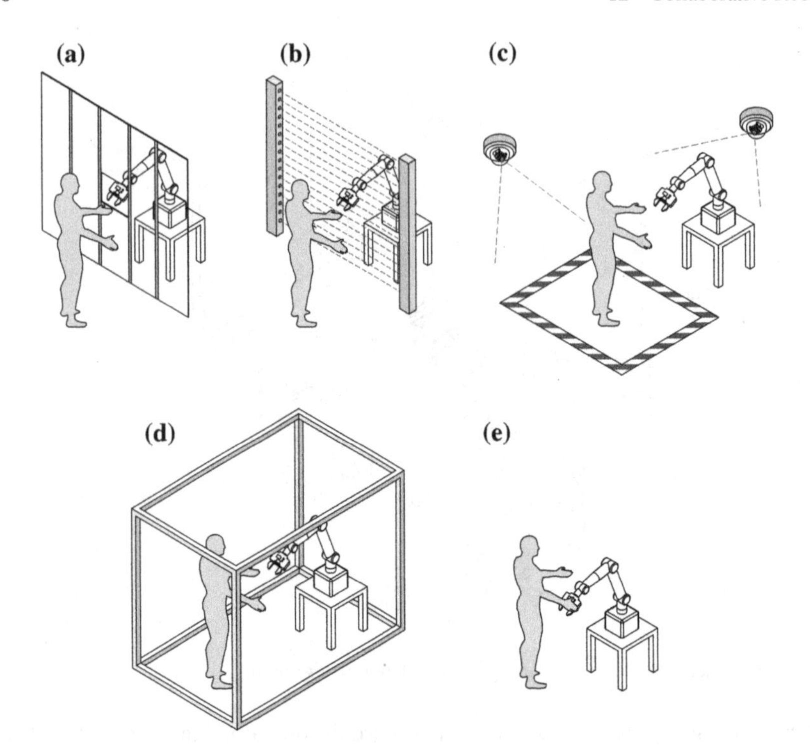

Fig. 12.10 Conceptual applications of collaborative robots: **a** hand-over window, **b** interface window, **c** collaborative workspace, **d** inspection, and **e** hand-guided robot (ISO 10218-2:2011)

sensitive guards around the workspace where the application is performed in automatic mode without limitations. Interaction with the operator is performed through a window. In the vicinity of the window the robot reduces its speed. The window also acts as the limit for the robot workspace.

The interface window (Fig. 12.10b) acts as a barrier for the robot system. On the robot side the robot can perform autonomous automatic operations. The robot system is also guarded by fixed or sensitive guards around the workspace. The robot stops at the interface window and can be then manually moved outside the interface. For guided movement the robot must be equipped with hand guiding device. This method is used for automatic stacking, guided assembly, guided filling, testing, benching, and cleaning.

Applications including simple assembling and handling can take advantages of the collaborative workspace (Fig. 12.10c). Inside the common workspace the robot can perform automatic operations. When the operator enters the collaborative workspace, the robot reduces speed and/or stops. In this type of application, additional person-detection systems using one or more sensors are needed.

Applications including inspection and parameters tuning (e.g., welding application, see Fig. 12.10d) require guarded workspace and person-detection systems.

When the operator enters the shared workspace, the robot continues operation with reduced speed. The application needs to have additional measures to prevent misuse.

Hand-guided robots (Fig. 12.10e) are used for hand-guided applications (e.g., assembling or painting). The robot is equipped with hand-guiding device. The operator guides the robot by hand along a path in a task-specific workspace with reduced speed. The area of collaborative workspace is mainly dependent on the hazards of the required application.

Chapter 13
Mobile Robots

A mobile robot is a device that is capable of locomotion. It has the ability to move around its environment using wheels, tracks, legs, or a combination of them. It may also fly, swim, crawl, or roll. Mobile robots are used for various applications in factories (automated guided vehicles), homes (floor cleaning devices), hospitals (transportation of food and medications), in agriculture (fruit and vegetable picking, fertilization, planting), for military as well as search and rescue operations. They address the demand for flexible material handling, the desire for robots to be able to operate on large structures, and the need for rapid reconfiguration of work areas.

Though mobile robots move in different ways, the focus in this chapter will be on devices that use wheels for locomotion (walking robots are presented in Chap. 14). In industrial applications automated guided vehicles (AGVs) are of special interest to move materials around a manufacturing facility or a warehouse. Tuggers typically pull carts (Fig. 13.1a), unit loaders use a flat platform to transport a unit load stacked on the platform (Fig. 13.1b), and mobile forklifts are used to automatically pickup and drop loads off from various heights (Fig. 13.1c). AGVs typically follow markers or wires in the floor, or use vision, magnets, or lasers for moving around the facility. This organized movement is called navigation; a process or activity to plan and direct a robot along a route or path to move safely from one location to another without getting lost or colliding with other objects.

Navigation is typically a complex task consisting of localization, path planning and motion control. Localization denotes robot's ability to establish its own position and orientation within the global coordinate frame. Autonomous path planning represents determination of a collision-free path for a robot between start and goal positions between obstacles cluttered in a workspace. This also includes interactions between mobile robots and humans and between groups of mobile robots. Motion control must guarantee execution of movement along the planned path with simultaneous obstacle avoidance.

In collaborative settings humans and robots share a workspace resulting in a need for improved human-robot communication and for robot awareness of people around it. The robot must typically keep a safe distance from people. However, devices like

M. Mihelj et al., *Robotics*, https://doi.org/10.1007/978-3-319-72911-4_13

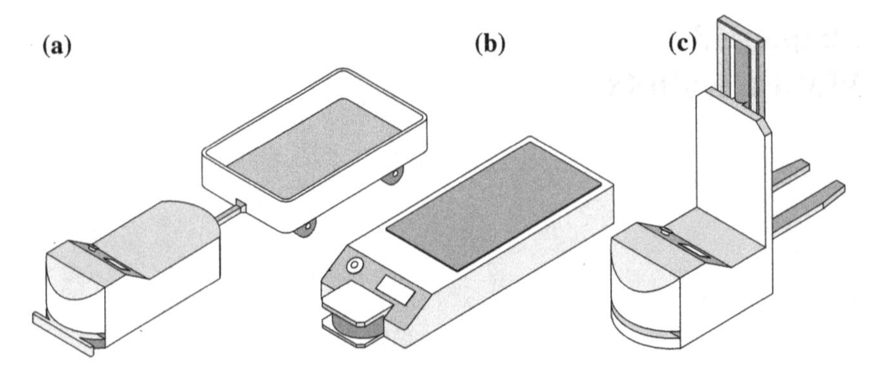

Fig. 13.1 Automated guided vehicles: **a** Tugger, **b** unit loader, and **c** mobile forklift

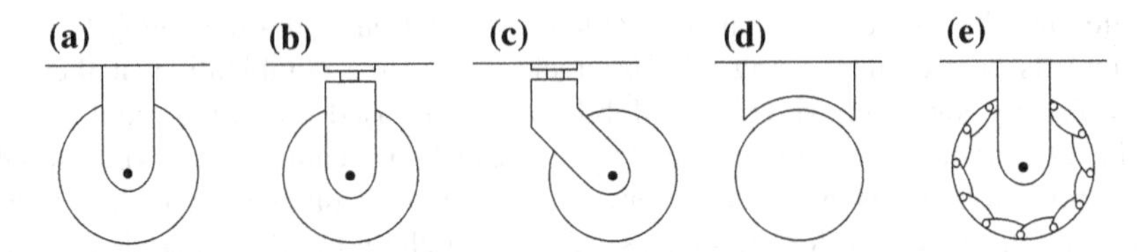

Fig. 13.2 Wheel designs: **a** Standard fixed wheel, **b** standard steered wheel, **c** castor wheel, **d** spherical wheel, and **e** Swedish wheel

personal care robots, require close proximity between the human and the robot and these machines are examples of advanced human-robot interactive systems.

13.1 Mobile Robot Kinematics

With its simple mechanical design, the wheel is the most popular locomotion mechanism in mobile robotics. Wheels provide traction and three wheels guarantee stable robot balance. Wheels can be designed in different forms as shown in Fig. 13.2.

The fixed wheel, the standard steered wheel and the castor wheel have a primary axis of rotation and are directional. Movement in different direction is not possible without first steering the wheel around the vertical axis. The spherical wheel is omnidirectional as it enables movement in all directions without steering first. The Swedish wheel tries to achieve omnidirectional behavior with passive rollers attached around the circumference of the wheel. Thus, the wheel can move along different trajectories, as well as forwards and backwards.

Selection of wheel type, number of wheels, as well as their attachment to the robot chassis significantly affect mobile robot kinematics. Examples of kinematic designs are shown in Fig. 13.3. They range from two-wheel to four-wheel configurations. The two platforms in the righthand column are omnidirectional.

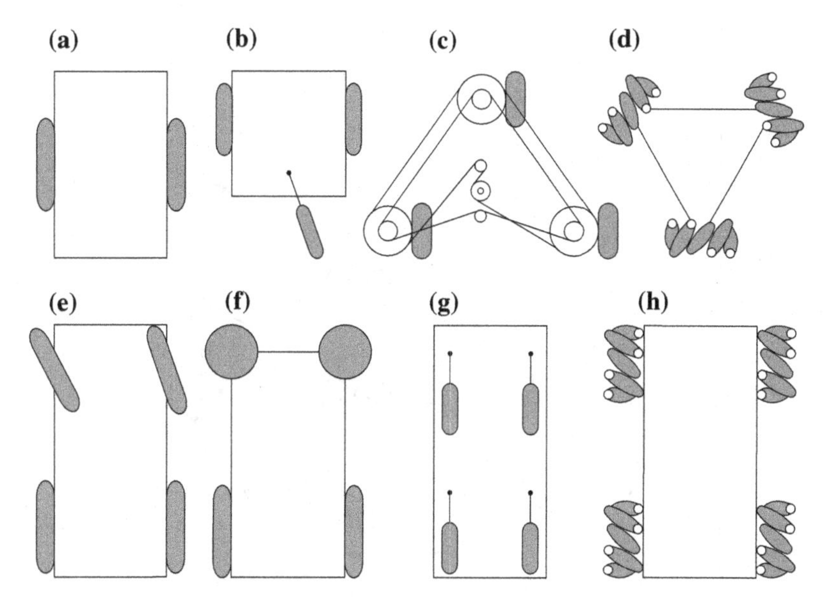

Fig. 13.3 Mobile robot configuration examples: **a** two-wheel differential drive, **b** differential drive with castor wheel, **c** three synchronously motorized and steered wheels, **d** three omnidirectional wheels in triangle, **e** four wheels with car-like steering, **f** two differential traction wheels and two omnidirectional wheels, **g** four motorized and steered castor wheels, and **h** four omnidirectional wheels in rectangular configuration

For the purpose of analysis, a mobile robot will be represented as a rigid body on wheels that can move only in a horizontal plane. With these assumptions the pose of the robot can be defined with three coordinates, two representing position in the horizontal plane and one describing orientation around the vertical axis. Relations are presented in Fig. 13.4 for a simple differential drive mechanism. Axes x_G and y_G define the global coordinate frame. The robot local coordinate frame is defined with axes x_m and y_m. The x_m axis points in the robot forward direction.

Robot position and orientation are defined with the following vector

$$\mathbf{x} = \begin{bmatrix} x \\ y \\ \varphi \end{bmatrix}, \tag{13.1}$$

where x and y coordinates define robot position relative to the global coordinate frame and angle φ determines its orientation (rotation around vertical axis). Robot orientation can be described also in the form of a rotation matrix

$$\mathbf{R} = \begin{bmatrix} \cos\varphi & -\sin\varphi & 0 \\ \sin\varphi & \cos\varphi & 0 \\ 0 & 0 & 1 \end{bmatrix}. \tag{13.2}$$

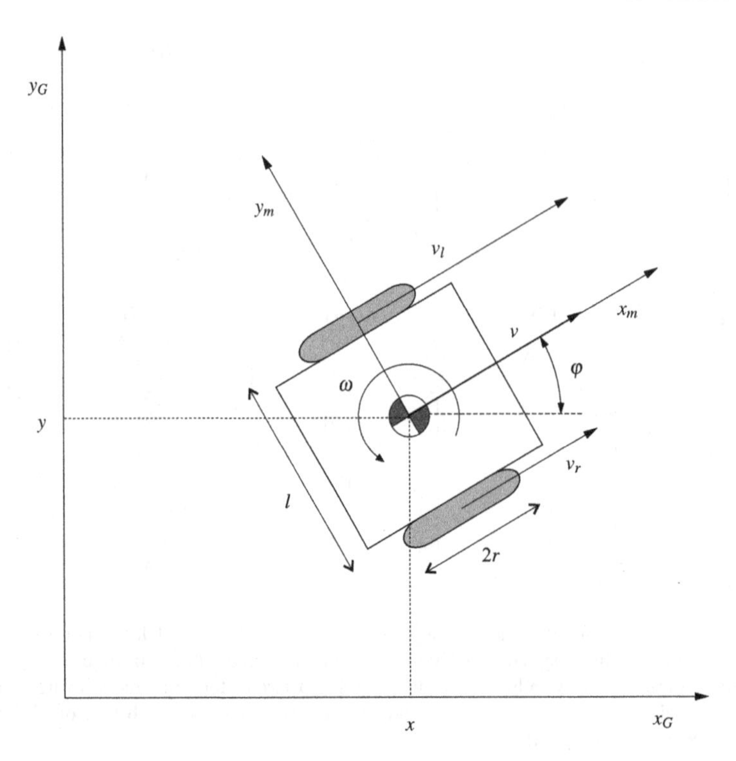

Fig. 13.4 Position and orientation of a mobile robot—differential drive robot example

Homogenous transformation matrix describing the pose of the mobile robot is then

$$\mathbf{T} = \begin{bmatrix} \cos\varphi & -\sin\varphi & 0 & x \\ \sin\varphi & \cos\varphi & 0 & y \\ 0 & 0 & 1 & 0 \\ 0 & 0 & 0 & 1 \end{bmatrix}. \tag{13.3}$$

The differential drive robot presented in Fig. 13.4 has a simple mechanical structure. Its movement is based on two separately driven wheels attached on either side of the robot body. The robot changes its direction by varying the relative speed of rotation of its wheels. Thus, it does not require an additional steering motion. If wheels are driven in the same direction and with equal speed, the robot will follow a straight line. If wheels are turned with equal speed in opposite directions, the robot will rotate about the middle point between the wheels. In general, the center of robot rotation may lay anywhere on the line through wheel axes and will depend on each wheel speed of rotation and its direction.

With its simple kinematics it is an ideal model for studying robot movement. By representing robot width (distance between tire contact points with the ground) with l and wheel radius with r the robot motion can be analyzed. The wheels rotate with

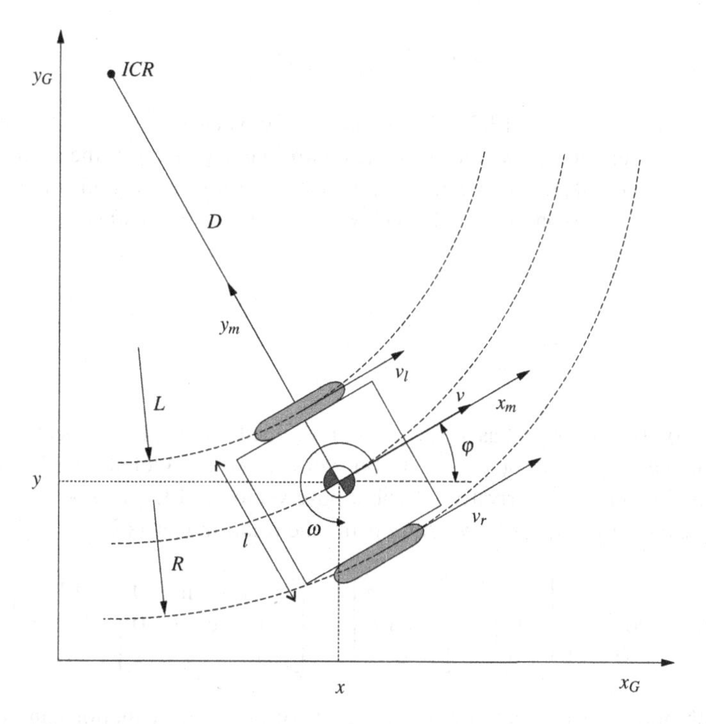

Fig. 13.5 Differential drive robot kinematics

angular rates ω_r (right wheel) and ω_l (left wheel), resulting in wheel speeds v_r and v_l of the right and left wheel, respectively

$$\begin{aligned} v_r &= \omega_r r, \\ v_l &= \omega_l r. \end{aligned} \tag{13.4}$$

The two wheel rotations result in the robot translational speed along robot x_m axis and angular rate around its vertical axis. With reference to Fig. 13.5 the angular rate can be defined as

$$\omega = \frac{v_l}{D - \frac{l}{2}} = \frac{v_r}{D + \frac{l}{2}}, \tag{13.5}$$

where D is the distance between the middle point on the robot (in this case the origin of the frame x_m–y_m) and the point that defines the instantaneous center of rotation (*ICR*). The *ICR* is the point in the horizontal plane around which the robot rotates at a specific instant of time. From equality in (13.5) the following relation can be derived

$$\omega = \frac{v_r - v_l}{l} = \frac{r}{l}(\omega_r - \omega_l). \tag{13.6}$$

Translational speed along the x_m axis can then be determined as

$$v = \omega D = \frac{v_r + v_l}{2} = \frac{r}{2}(\omega_r + \omega_l). \tag{13.7}$$

Equations (13.6) and (13.7) define relations between wheels' angular rates and mobile robot velocity. However, from the control perspective it is the more relevant inverse relation that defines wheels' angular rates from the desired robot velocity. By combining (13.6) and (13.7) the following relations are obtained

$$\begin{aligned} \omega_r &= \frac{2v + \omega l}{2r}, \\ \omega_l &= \frac{2v - \omega l}{2r}. \end{aligned} \tag{13.8}$$

Robot velocity determined as a pair $[v, \ \omega]$ is defined relative to the local coordinate frame of the mobile robot x_m–y_m. Robot velocity in the global coordinate frame x_G–y_G defined as time derivative of robot pose vector $\mathbf{x}$ (13.1) can be computed by rotating the locally expressed velocity using the rotation matrix $\mathbf{R}$ (13.2) as

$$\begin{bmatrix} \cos\varphi & -\sin\varphi & 0 \\ \sin\varphi & \cos\varphi & 0 \\ 0 & 0 & 1 \end{bmatrix} \begin{bmatrix} v \\ 0 \\ 0 \end{bmatrix} = \begin{bmatrix} v\cos\varphi \\ v\sin\varphi \\ 0 \end{bmatrix}, \quad \begin{bmatrix} \cos\varphi & -\sin\varphi & 0 \\ \sin\varphi & \cos\varphi & 0 \\ 0 & 0 & 1 \end{bmatrix} \begin{bmatrix} 0 \\ 0 \\ \omega \end{bmatrix} = \begin{bmatrix} 0 \\ 0 \\ \omega \end{bmatrix}. \tag{13.9}$$

By combining translational and rotation parts of the above equations and omitting elements that are zero, the mobile robot velocity in the global coordinate frame can be written as

$$\dot{\mathbf{x}} = \begin{bmatrix} \dot{x} \\ \dot{y} \\ \dot{\varphi} \end{bmatrix} = \begin{bmatrix} v\cos\varphi \\ v\sin\varphi \\ \omega \end{bmatrix}. \tag{13.10}$$

From Eq. (13.10) it is clear that relevant quantities for describing mobile robot movement are translational velocity along robot x_m axis v, rotational velocity around vertical axis ω, and robot orientation with respect to the global coordinate frame φ. With this in mind we may further simplify the differential drive robot into a unicycle model (as shown in Fig. 13.6). Now the above-mentioned three quantities describe the movement of the unicycle represented as a single wheel with marked forward direction in the middle of the differential drive robot in Fig. 13.6. The unicycle can be easily transformed back to the differential drive robot based on Eq. (13.8).

The attractive property of the unicycle model is its simplicity. Therefore, it will be used throughout this chapter for analysis. However, the model can be in general converted back to any other kinematically more complex mobile robot. As an example, we review a mobile platform based on the car steering principle shown in Fig. 13.7.

The car steering geometry solves the problem of wheels on the inside and outside of a turn needing to trace circles of different radii. Therefore, steering angles of left and right front wheels are different. In the unicycle model the orientation of the unicycle is defined with angle φ, the same as the orientation of the differential drive robot.

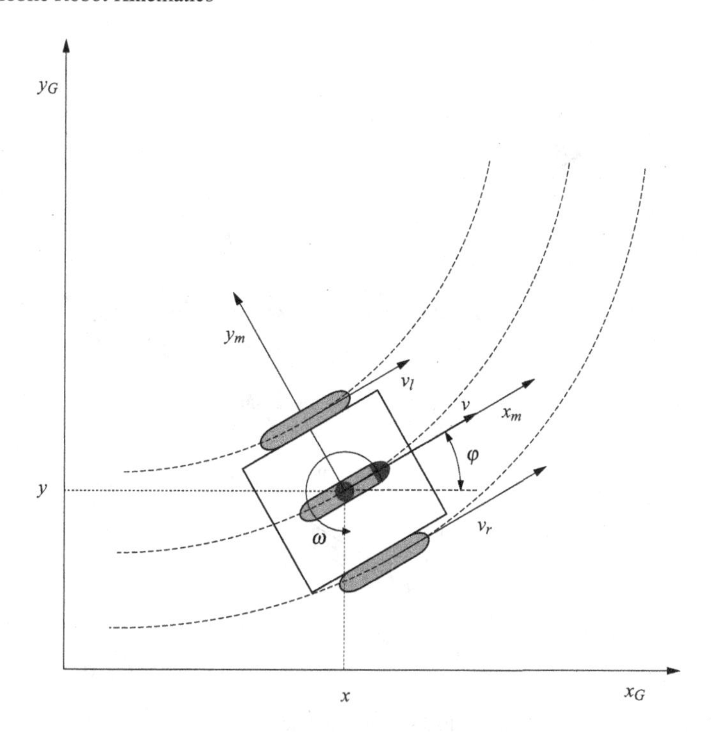

Fig. 13.6 Unicycle model of a differential drive mobile robot

In the car-like problem the orientation of the mobile robot is defined by angle φ. The unicycle model is positioned in the middle of the front wheels and its orientation is defined such to achieve the same instantaneous center of rotation as defined by the orientation of the car's left and right wheels. The unicycle is now the third front wheel and the *ICR* is positioned at the intersection point of all the three lines perpendicular to the front wheels. Angle ψ is now defined as the deviation of the unicycle orientation from the robot x_m axis (as shown in Fig. 13.7). By computing angle ψ the relation between the car-like robot and the unicycle will be established.

By following the same principle as in (13.7), translational velocity of the unicycle can be defined as

$$v = D\omega, \tag{13.11}$$

where D is the distance between the unicycle and the *ICR*. Distance D can then be computed as

$$D = \frac{v}{\omega}. \tag{13.12}$$

Path curvature for the unicycle $\mathscr{K}_u$ can be defined as the inverse of the instantaneous radius of rotation as

$$\mathscr{K}_u = \frac{1}{D} = \frac{\omega}{v}. \tag{13.13}$$

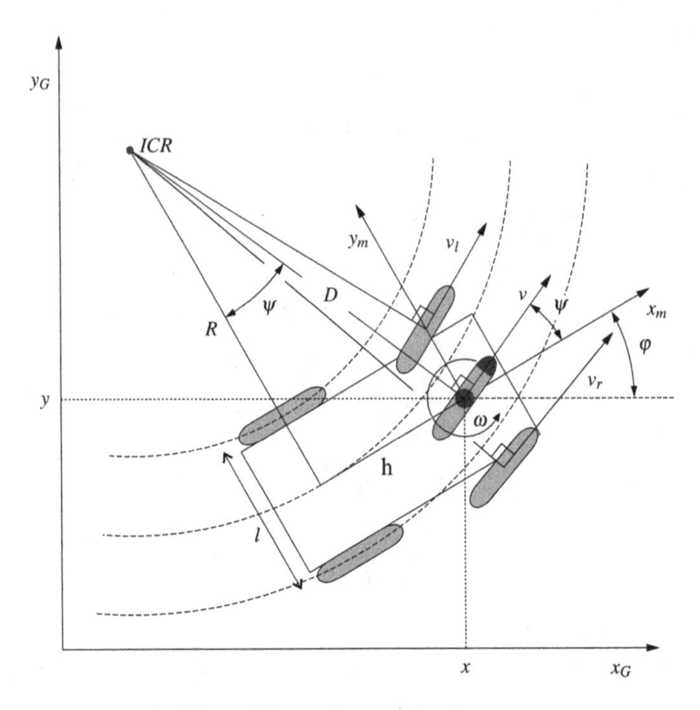

Fig. 13.7 Unicycle model of the car-like steering mobile robot

By considering car kinematics, the following relation can be written from Fig. 13.7.

$$h = D \sin \psi, \tag{13.14}$$

where angle ψ is also the angle between lines D and R (the distance between ICR and the middle point between the rear wheels of the vehicle) and h is the distance between the center of the unicycle and the middle point between the rear wheels of the robot. Distance D can then be computed as

$$D = \frac{h}{\sin \psi} \tag{13.15}$$

and the curvature for the car $\mathscr{K}_c$ is then defined as

$$\mathscr{K}_c = \frac{1}{D} = \frac{\sin \psi}{h}. \tag{13.16}$$

With equal $\mathscr{K}_c$ and $\mathscr{K}_u$ the following relation can be obtained

$$\mathscr{K}_c = \mathscr{K}_u \quad \Rightarrow \quad \sin \psi = \frac{\omega l}{v}. \tag{13.17}$$

Finally, the angle ψ equals

$$\psi = \arcsin \frac{\omega l}{v}. \tag{13.18}$$

Angle ψ is the desired steering angle for the car and it can be computed from the known speed v, angular rate ω, and width of the car l.

With the defined relation between the unicycle and a mobile robot with other kinematics the analysis can be based on a simple unicycle model and generalized to the other robot.

13.2 Navigation

Mobile robots often operate in unknown and unstructured environments and need to self-localize, plan a path to a goal, build and interpret the map of the environment, and then control their motion through that environment.

13.2.1 Localization

An important difference between a manipulator and a mobile robot is in position estimation. A manipulator has a fixed base and by measuring robot joint positions and knowing its kinematic model it is possible to determine the pose of its end-effector. A mobile robot can move as one unit through the environment and there is no direct way for measuring its position and orientation. A general solution is to estimate the robot position and orientation through integration of motion (velocity) over time.

However, more accurate and often also more complex approaches are typically required. If the map of the environment is known in advance mobile robot paths can be preplanned. This is specifically useful when the environment is relatively static and robust operation is required, such as in industrial applications. More complex approaches are based on dynamic path planning based on sensor information and recognition of features in the environment. The robot first determines its own position and plans its movement through traversable areas. When the workspace or the tasks change frequently it is typically better to plan dynamically. Often a trade-off is required between preplanning and dynamic generation of plans. In order to simplify the task, markers may be placed in the environment. These markers can be easily recognized by sensors on the robot and provide accurate localization.

Automated guided vehicles in industrial environments make use of various navigation/guidance technologies: magnetic tape, wire, magnetic spot, laser, and natural.

Localization and path planning are often based on electrified wires embedded in the floor using inductive guidance. A guide path sensor is mounted on the vehicle. The wire can be replaced by magnetic tape or a painted line (Fig. 13.8a). In the latter

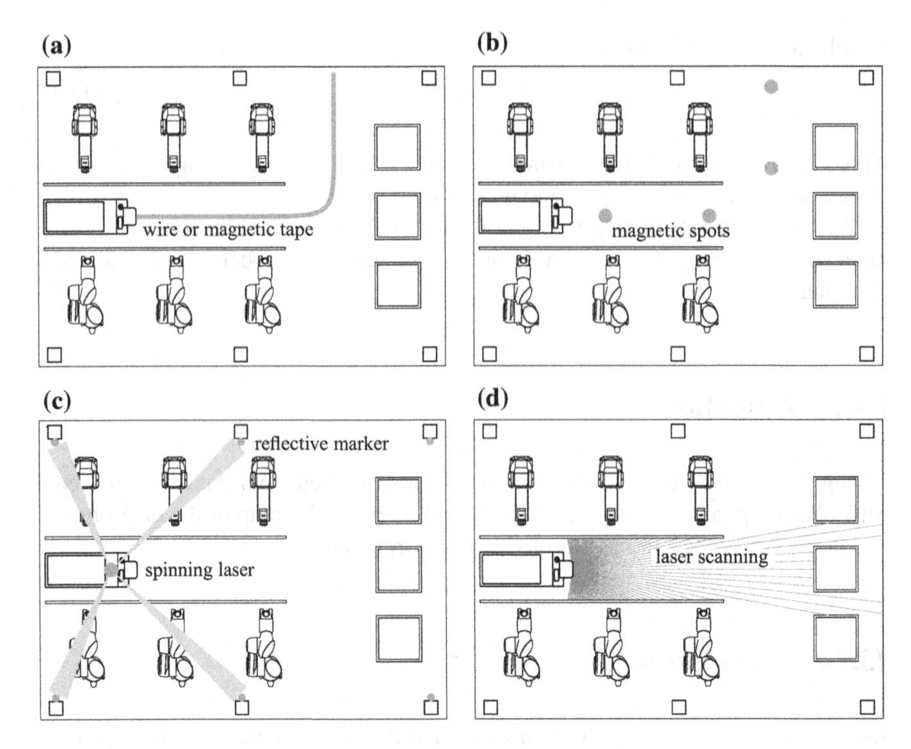

Fig. 13.8 Sensor abstraction disk from the suit of sensors on board the robot

case the robot uses a camera to determine its relative position to the floor line. Paths are fixed and continuous. Unique markers may be placed along the line to indicate specific positions. Instead of placing lines and markers on the floor, markers (two-dimensional patterns) can also be put on the ceiling to be identified by an onboard camera. Magnetic spot guidance uses path marked with magnetic pucks (Fig. 13.8b). Paths are open and changeable.

Floor-based localization techniques are often replaced by laser-based methods. Laser triangulation methods, in which a spinning laser senses range and azimuth to wall-mounted reflectors, provide accurate localization information without the need to follow specific lines on the floor. Laser guidance technology uses multiple, fixed reference points (reflective strips) located within the operating area that can be detected by a laser head mounted on the vehicle (Fig. 13.8c). As the facility is mapped in advance, paths can be easily changed and expanded.

Natural navigation is based on information of the existing environment scanned by laser scanners, with the aid of a few fixed reference points (Fig. 13.8d). Area is mapped in advance. Natural navigation is flexible and expendable. It is suitable for environments that change frequently but not significantly. In confined spaces the robot may follow the wall through the environment range-basing from the wall.

Radio-based indoor positioning systems are also being introduced that enable robot localization in a similar manner as the outdoor global positioning system. Localization is based on triangulation with fixed beacons mounted in the facility and the sensor mounted on the robot. Distances are computed by measuring travel time of radio waves from the beacon to the sensor.

13.2.1.1 Odometry

A simple and commonly-used approach for robot localization is to rely on odometry, which uses information from motion sensors (typically wheel encoders) to estimate change in position over time. These position changes are accumulated using integration principles providing the robot position relative to a starting location. The method is sensitive to errors due to integration of velocity measurements over time to give position estimates.

Analysis of robot motion starts with the understanding of the contribution of each wheel to the velocity of the robot. For the specific case of a differential drive robot these relations are defined in (13.6) and (13.7). Wheel speed may be directly measured using a tachometer. If such a sensor is not available, the speed can be estimated through numerical differentiation of the position obtained from encoders. In such case speeds for the right and left wheel can be computed as

$$\begin{aligned} v_r &= 2\pi r \frac{n_r(t) - n_r(t - \Delta t)}{N \Delta t}, \\ v_l &= 2\pi r \frac{n_l(t) - n_l(t - \Delta t)}{N \Delta t}, \end{aligned} \tag{13.19}$$

where r is the wheel radius, N is the encoder resolution in terms of counts per revolution, n_r and n_l are encoder counts of the right and left wheel at time t, respectively, and $n_r(t - \Delta t)$ and $n_l(t - \Delta t)$ are the same quantities at the previous sampling time.

Robot position and orientation can then be estimated with numerical integration of Eq. (13.10) and consideration of (13.6) and (13.7) as

$$\begin{aligned} x(t) &= x(t - \Delta t) + v \cos\varphi \Delta t = x(t - \Delta t) + \frac{v_r + v_l}{2} \cos\varphi \Delta t, \\ y(t) &= y(t - \Delta t) + v \sin\varphi \Delta t = x(t - \Delta t) + \frac{v_r + v_l}{2} \sin\varphi \Delta t, \\ \varphi(t) &= \varphi(t - \Delta t) + \omega \Delta t = \varphi(t - \Delta t) + \frac{v_r - v_l}{l} \Delta t. \end{aligned} \tag{13.20}$$

Different factors reduce the effectiveness of odometry-based methods for robot position estimation. A very important factor is wheel slippage that significantly reduces precision of position estimation. Performance may be improved by using models of the errors and of the vehicle. Floor spots or magnets may be used to correct for odometry errors that accumulate between these points. Odometry can also be augmented by sensor-based measurements from lasers, cameras, radiofrequency identification systems, and beacons.

13.2.1.2 Simultaneous Localization and Mapping

More advanced systems make use of algorithms that accomplish the navigation subtasks (localization, path planning) simultaneously. The approach that is concerned with the problem of building a map of an unknown environment by a mobile robot while at the same time navigating the environment using the map is called simultaneous localization and mapping (SLAM). By observing the same features in multiple views using sensors that move with the vehicle, the SLAM algorithm accumulates and combines together the sensor information. By combining the robot position estimation with the gathered information, a local map can be constructed by stitching together available data. Over time, the complete environment can be mapped and the maps can be used to plan the robot paths.

SLAM consists of multiple parts, such as landmark extraction, data association, state estimation, state update and landmark update. There are many ways to solve each of the smaller parts, but they are beyond the scope of this book.

13.2.1.3 Sensor Abstraction Disk

When the mobile robot is moving through the environment it must also observe its surroundings. Sensors on-board the robot look for obstacles or unexpected objects in the path of the vehicle and the robot may be able to plan a way around them before returning to the pre-planned route. A typical suite of sensors includes infrared proximity sensors, ultrasonic distance sensors, laser scanners, vision, tactile sensing, and global positioning sensors. Sensors are strategically placed onboard the robot and around its circumference. Each sensor provides different information in terms of quantity, quality, range, and resolution. However, typically information from all sensors is combined to provide an accurate image of the robot environment. Without dealing specifically with analysis of individual sensors and integration of sensory information it is possible to assume that distance and direction to all obstacles from the robot's perspective can be obtained from the sensor suite. The sensor abstraction disk presented in Fig. 13.9 is an example of sensory integration providing information about obstacles within the radius of the disk around the robot.

From the known position d_o and orientation φ_o of the obstacle and the known pose of the robot $[x, y, \varphi]^T$, it is possible to determine the obstacle position (x_o, y_o) in the global coordinate frame as

$$\begin{aligned} x_o &= x + d_o \cos(\varphi + \varphi_o), \\ y_o &= y + d_o \sin(\varphi + \varphi_o). \end{aligned} \tag{13.21}$$

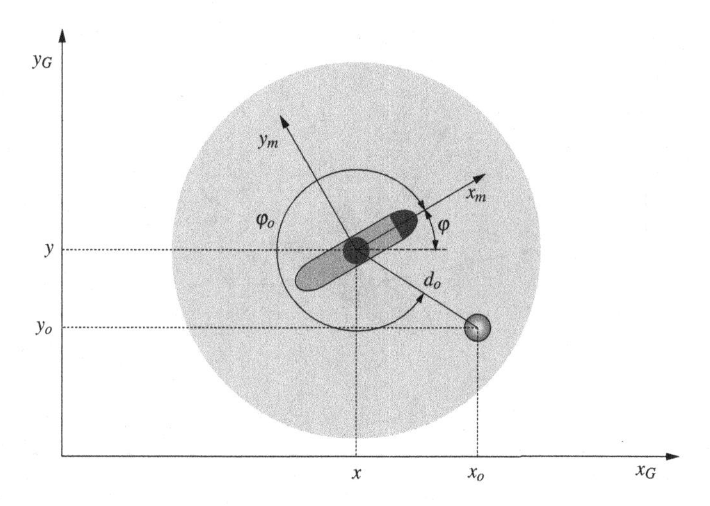

Fig. 13.9 Sensor abstraction disk from the suit of sensors on board the robot

The following analysis will be based on the assumptions of a unicycle robot model and the information about objects obtained from the sensor abstraction disk.

13.2.2 Path Planning

Path planning enables autonomous mobile robots to track an optimal collision free path from the starting position to the goal without colliding with obstacles in the workspace area. An ideal path planner must be able to handle uncertainties in the sensed world model, to minimize the impact of objects to the robot, and to find the optimum path in minimum time especially if the path is to be negotiated regularly. In general, the path planning should result in the path with the lowest possible cost, it should be fast and robust as well as generic with respect to different maps.

Different algorithms are available for (real-time) path planning. A simple method consists of combining straight-line segments connected with vertices. Another standard search method for finding the optimal path is the A* algorithm with its modifications. The algorithm finds a directed path between multiple points, called nodes. The robot environment represented with a map can be decomposed into free and occupied spaces. Then A* search can be performed to find a piecewise linear path through the free nodes.

An artificial potential field algorithm can be used for obstacle avoidance. The algorithm uses repulsive potential fields around the obstacles to force away the robot subjected to this potential and use an attractive potential field around the goal to attract the robot to go to the goal. Repulsive and attractive fields modify the robot's path. The algorithm enables real-time operations of a mobile robot in a complex environment.

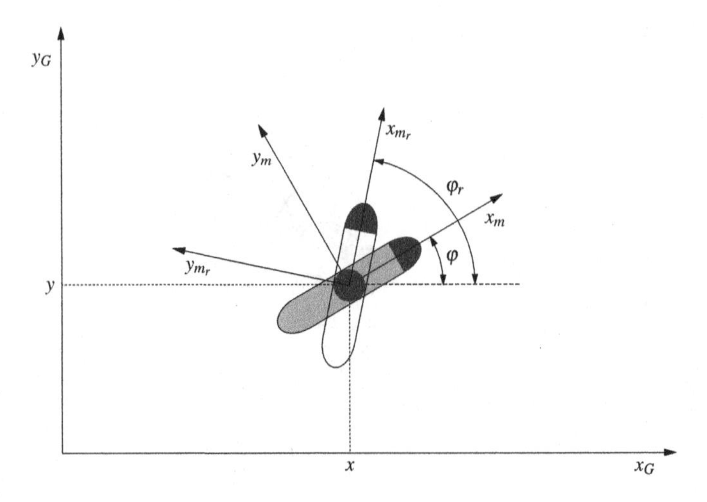

Fig. 13.10 Unicycle orientation control; grey unicycle represents actual robot and white unicycle represents desired orientation

13.2.3 Path Control

In order to complete the task, the mobile robot needs to move from its initial location to the desired final position and orientation. A control system is required to control the vehicle along its path.

13.2.3.1 Control of Orientation

Based on the unicycle model presented in Fig. 13.10 control of orientation will first be considered. A similar approach would be valid for mobile robots that can change orientation without changing their position (a differential drive robot is such a vehicle, but the car is not).

The control goal is to minimize the orientation error

$$\tilde{\varphi} = \varphi_r - \varphi, \tag{13.22}$$

where φ_r is the desired orientation and φ is the actual orientation. We assume that the control is based on proportional-integral-derivative (PID) control approach

$$PID(\tilde{\varphi}) = K_p\tilde{\varphi} + K_i \int \tilde{\varphi}dt + K_d\dot{\tilde{\varphi}} \tag{13.23}$$

or one of its subversions, such as proportional-derivative controller. Then the desired angular velocity of the mobile robot can be computed as

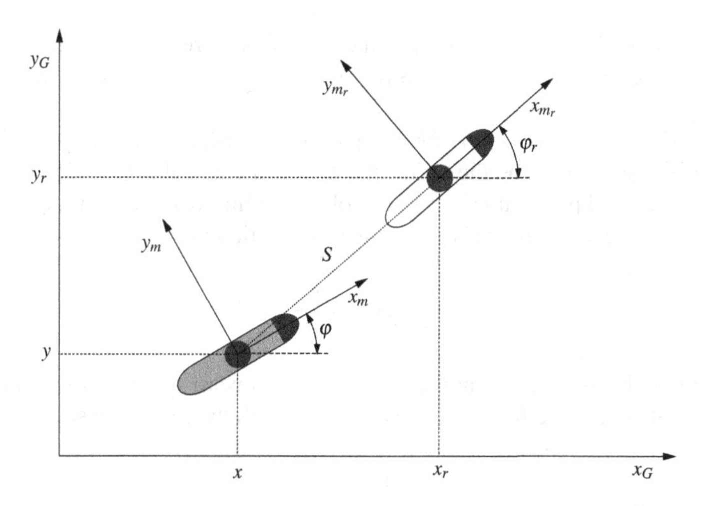

Fig. 13.11 Unicycle position and orientation control; grey unicycle represents actual robot and white unicycle represents goal location

$$\omega = K_p\tilde{\varphi} + K_i \int \tilde{\varphi} dt + K_d\dot{\tilde{\varphi}}. \tag{13.24}$$

It should be noted that angles are periodic functions and if we assume configuration

$$\varphi_r = 0 \quad \wedge \quad \varphi = 2\pi \quad \Rightarrow \quad \tilde{\varphi} = -2\pi, \tag{13.25}$$

the robot will spin once before it will reach the final orientation. This is usually not desirable robot behavior. Therefore, orientation error must be limited such to require at maximum π radians rotation in either direction

$$\tilde{\varphi} \in [-\pi, \pi]. \tag{13.26}$$

A simple solution is to use a four-quadrant arctan function as

$$\tilde{\varphi} = \arctan(\sin\tilde{\varphi}, \cos\tilde{\varphi}) \in [-\pi, \pi]. \tag{13.27}$$

With the combination of (13.27) and (13.24) the robot will reach the desired orientation without rotating more than half circle in positive or negative direction.

13.2.3.2 Control of Position and Orientation

The mobile robot typically moves from its initial location to its final (goal) location which requires change of position and orientation. Since the robot needs to move to its goal location we will refer to this task as *go-to-goal*. Figure 13.11 represents such conditions. Coordinate frame x_m–y_m defines the robot current pose and frame x_{m_r}–y_{m_r} defines the goal pose. Line segment S represents the shortest path for completing the task.

The desired robot orientation for completing the task can be defined as the angle between line segment S and the horizontal axis of the global coordinate frame. With the known desired position $(x_r,\ y_r)$ and robot current position $(x,\ y)$, angle φ_r can be computed at every time instant during robot motion as

$$\varphi_r = \arctan \frac{y_r - y}{x_r - x}. \tag{13.28}$$

By assuming that the robot is moving at constant forward speed v_0, robot movement in the global coordinate frame can be described with the following set of equations

$$\begin{aligned} \dot{x} &= v_0 \cos \varphi, \\ \dot{y} &= v_0 \sin \varphi, \\ \dot{\varphi} &= \omega = PID(\tilde{\varphi}). \end{aligned} \tag{13.29}$$

With this approach the control goal is to maintain constant speed v_0 and track the desired angle φ_r computed from (13.28). If we assume a differential drive robot, wheel angular rates can then be computed from (13.8) as

$$\begin{aligned} \omega_r &= \frac{2v_0 + \omega l}{2r}, \\ \omega_l &= \frac{2v_0 - \omega l}{2r}. \end{aligned} \tag{13.30}$$

When moving with constant velocity v_0 the robot would overshoot its goal location. Therefore, it is reasonable to define robot forward speed based on the distance to the goal

$$G = \sqrt{(x_r - x)^2 + (y_r - y)^2}. \tag{13.31}$$

With a proportional controller, the desired speed can be defined as

$$v_G = K_v G, \tag{13.32}$$

where K_v is the velocity gain. Equations (13.29) can then be rewritten as

$$\begin{aligned} \dot{x} &= v_G \cos \varphi, \\ \dot{y} &= v_G \sin \varphi, \\ \dot{\varphi} &= \omega = PID(\tilde{\varphi}) \end{aligned} \tag{13.33}$$

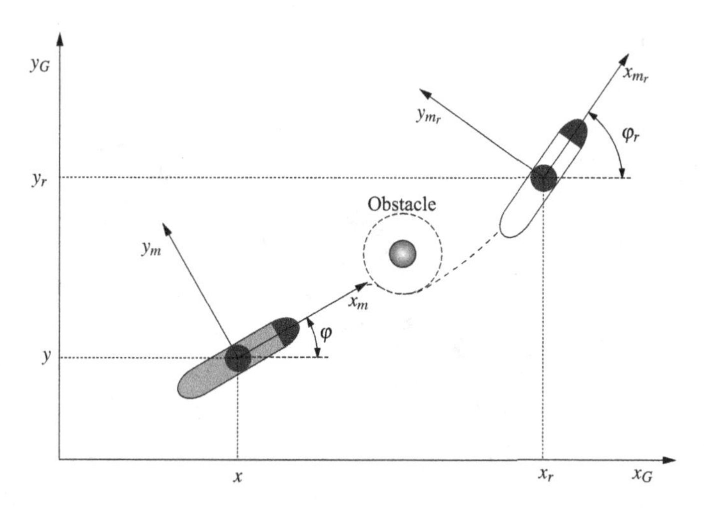

Fig. 13.12 Unicycle position and orientation control with obstacle avoidance; grey unicycle represents actual robot and white unicycle represents goal location; gray circle is the obstacle and dashed circular line is safe zone around the obstacle

and in (13.30) v_0 must be replaced by v_G. With this approach the robot will decelerate when approaching the goal location. Since desired speed increases with the distance to the goal, a maximum limit can be set on $v_G \in [0, v_{G_{max}}]$.

13.2.3.3 Obstacle Avoidance

Figure 13.12 shows conditions with an obstacle in the robot's path to the goal position. The robot cannot proceed directly to the target location without first avoiding the obstacle. Based on the concept of the sensor abstraction disk we assume that the robot is capable of detecting and locating the obstacle from a safe distance and using this information, can plan avoidance activities. The obstacle in Fig. 13.12 is represented by a gray circle and the dashed circular line around the obstacle represents a safe zone around the obstacle. The robot would not be allowed to enter the dashed circle.

With this in mind, we now have two control objectives. The first is *go-to-goal* and the second is *avoid-obstacle*. A more detailed representation of the two control objectives is shown in Fig. 13.13, where d_o indicates distance from the robot to the obstacle, u_g is the control variable associated with the *go-to-goal* objective and u_o is the control variable associated with *avoid-obstacle* objective. In order to successfully complete the task, the u_g needs to point to the goal while the u_o needs to point away from the obstacle. The actual control variable u is the result of blending u_g and u_o.

The *go-to-goal* control part can be defined based on the distance to the goal position as

$$\begin{bmatrix} u_{g_x} \\ u_{g_y} \end{bmatrix} = K_g \begin{bmatrix} x_r - x \\ y_r - y \end{bmatrix}. \tag{13.34}$$

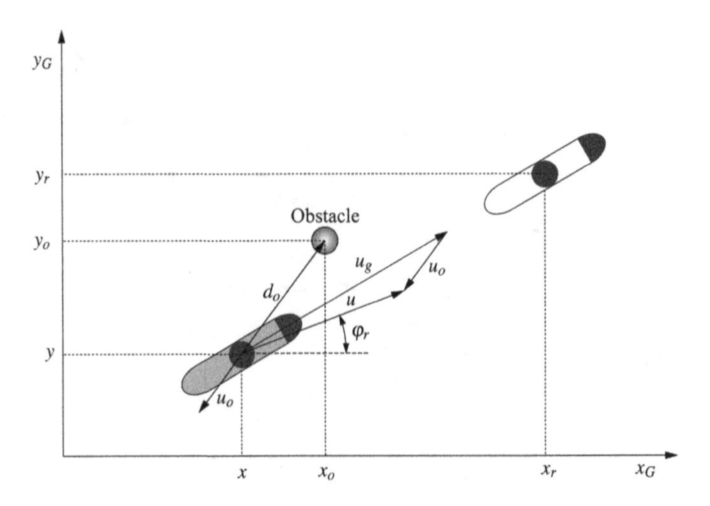

Fig. 13.13 Unicycle obstacle avoidance; grey unicycle represents actual robot, white unicycle represents goal location and grey circle is obstacle

Similarly, the *avoid-obstacle* control variable can be defined based on the distance to the obstacle

$$\begin{bmatrix} u_{o_x} \\ u_{o_y} \end{bmatrix} = K_o \begin{bmatrix} x - x_o \\ y - y_o \end{bmatrix}. \tag{13.35}$$

It should be noted that u_g points to the goal and u_o points away from the obstacle as seen by the definition of distances in the above two equations. Blending of the two control variables must be made based on the distance to the obstacle, which is defined as

$$\|d_o\| = \sqrt{(x_o - x)^2 + (y_o - y)^2}. \tag{13.36}$$

When the robot is far away from the obstacle, it only needs to proceed directly to the goal. However, in the vicinity of the obstacle the primary task becomes obstacle avoidance. Consecutively, blending can be implemented as

$$\begin{bmatrix} u_x \\ u_y \end{bmatrix} = \lambda(\|d_o\|) \begin{bmatrix} u_{g_x} \\ u_{g_y} \end{bmatrix} + (1 - \lambda(\|d_o\|)) \begin{bmatrix} u_{o_x} \\ u_{o_y} \end{bmatrix}, \quad \lambda(\|d_o\|) \in [0, 1]. \tag{13.37}$$

Parameter λ can, for example, be defined as an exponential function based on distance to the obstacle $\lambda = 1 - e^{-\kappa\|d_o\|}$ and parameter κ defines convergence rate of the function toward 1. As seen from Fig. 13.13 control variable u defines desired robot velocities in the global coordinate frame

$$\begin{bmatrix} \dot{x} \\ \dot{y} \end{bmatrix} = \begin{bmatrix} u_x \\ u_y \end{bmatrix}. \tag{13.38}$$

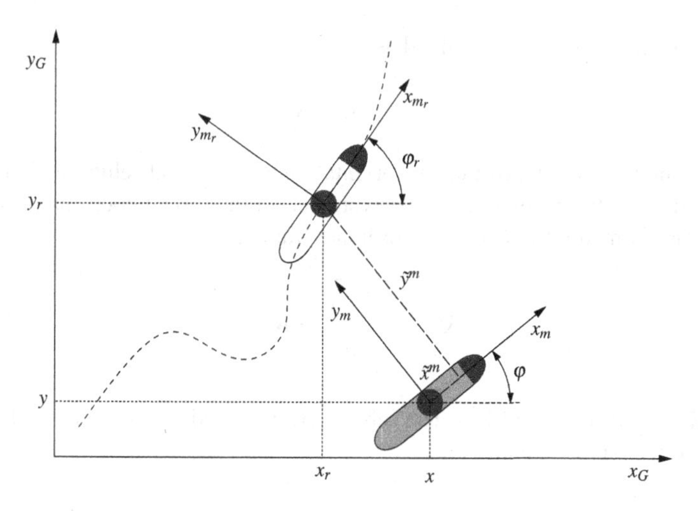

Fig. 13.14 Unicycle path following control; grey unicycle represents actual robot and white unicycle represents virtual vehicle on the path

Desired robot orientation can then be computed as

$$\varphi_r = \arctan \frac{u_y}{u_x}, \tag{13.39}$$

resulting in angular rate

$$\dot{\varphi} = \omega = PID(\tilde{\varphi}). \tag{13.40}$$

The forward robot speed can be computed as

$$v = \sqrt{\dot{x}^2 + \dot{y}^2} = \sqrt{v^2 \cos^2 \varphi + v^2 \sin^2 \varphi} = \sqrt{u_x^2 + u_y^2}. \tag{13.41}$$

Again, by assuming a differential drive robot, wheel angular rates can be computed from (13.8).

13.2.3.4 Path Following

Often the robot cannot just take the shortest path to the goal and it must follow a predefined path. In this case the control goal is to stay on the path. The task can be simplified by considering a virtual vehicle that moves along the path with a predefined speed. Then the control goal becomes tracking the virtual vehicle as shown in Fig. 13.14.

The tracking error can be defined as

$$\tilde{\mathbf{x}} = \mathbf{x}_r - \mathbf{x}, \tag{13.42}$$

where $\mathbf{x}_r$ and $\mathbf{x}$ represent position and orientation of the virtual vehicle and the mobile robot, respectively. All quantities are expressed in the global coordinate frame and can be transformed into the robot coordinate frame as

$$\tilde{\mathbf{x}}^m = \begin{bmatrix} \tilde{x}^m \\ \tilde{y}^m \\ \tilde{\varphi}^m \end{bmatrix} = \mathbf{R}^T \tilde{\mathbf{x}}, \tag{13.43}$$

where $\mathbf{R}$ is defined as in (13.2). The robot forward speed can be computed from the tracking error along x_m axis as

$$v = K_x \tilde{x}^m, \tag{13.44}$$

where K_x is the controller proportional gain. The angular rate must take into account the angle tracking error $\tilde{\varphi}^m = \tilde{\varphi}$ as well as distance to the path $\tilde{y}^m$. Namely, when the robot is away from the path it must steer toward the path. Thus, the control algorithm becomes

$$\omega = K_y \tilde{y}^m + K_\varphi \tilde{\varphi}^m, \tag{13.45}$$

where K_y and K_φ are controller proportional gains. Since velocity of the virtual vehicle is known (angular rate can be computed as the change of tangential direction along the path when the virtual vehicle moves forward), it can be taken into account as a feedforward control term. If v_r is the forward speed of the virtual vehicle and ω_r its angular rate, Eqs. (13.44) and (13.45) can be rewritten with the feedforward term as

$$v = v_r \cos\tilde{\varphi} + K_x \tilde{x}^m \tag{13.46}$$

and

$$\omega = \omega_r + K_y \tilde{y}^m + K_\varphi \tilde{\varphi}^m. \tag{13.47}$$

Chapter 14
Humanoid Robotics

Even before modern robotics began to develop, philosophers, engineers, and artists were interested in machines similar to humans. The first known example of a humanoid mechanism, which design has been preserved and can still be rebuilt today, is a mechanical knight created by Leonardo da Vinci and presented to the Milanese ruler Ludovico Sforza around 1495. The mechanism had a kinematic structure similar to present humanoid robots and it could move by a system of wires and pulleys. More recently writers like Karel Čapek and Isaac Asimov thought of robots that have a form similar to humans. There are several reasons why humanoid robots are thought to be interesting:

- Human environments are built for humans, therefore a general-purpose robot designed for human environments, e.g., homes, factories, hospitals, schools, etc., should have a form similar to humans to successfully operate in such environments.
- It is more natural for humans to interact and communicate with robots that look and behave in like humans.
- A humanoid robot can serve as an experimental tool to test the theories about human behavior created by computational neuroscientists, interested in how the human brain operates.

It can be said that modern humanoid robotics started with a series of humanoid robots created at the University of Waseda in Tokyo, Japan. The first of these robots was WABOT-1 created in 1973.

Despite recent progress in related areas such a soft robotics and artificial intelligence, humanoid robots that can operate in human-populated environments, where they collaborate and communicate with people in a natural way, are still only a distant dream. Currently, humanoid robots are at the stage where they can execute a variety of tasks. Tasks that are for example used in humanoid robot competitions, e.g. DARPA Robotics Challenge, include:

1. **Drive**: drive a utility vehicle down a lane blocked with barriers.
2. **Egress**: get out of the vehicle and locomote to a specified area.
3. **Door**: open a door and travel through a doorway.
4. **Valve**: turn a valve actuated by a hand-wheel.

M. Mihelj et al., *Robotics*, https://doi.org/10.1007/978-3-319-72911-4_14

5. **Wall**: use a tool (drill or saw) to cut through a concrete panel.
6. **Surprise** task, which was not known until the day of competition: remove a magnetic plug from one socket, insert it in a different socket.
7. **Rubble**: cross a debris field or negotiate irregular terrain.
8. **Stairs**: climb the stairs.

Modern humanoid robots can already execute such tasks autonomously, providing the approximate state of the environment is known in advance. However, it is still difficult for modern humanoid robots to perform such tasks without some prior information about the environmental conditions that can be exploited by a programmer to prepare the humanoid robot for the execution of multiple tasks. Integration and continuous sequencing of multiple robot actions remains a problem and some degree of teleoperation is still needed when performing longer task sequences.

While most of the standard robotics methodologies regarding robot kinematics, dynamics, control, trajectory planning, and sensing are relevant also when developing humanoid robots, humanoid robotics needs to deal with several specific issues. The foremost is the problem of biped locomotion and balance. Unlike other robots, humanoid robots must walk and keep balance during their operation. In the aforementioned robotics challenge, locomotion turned out to be one of the biggest issues. The basic indicator that describes the balance of a humanoid robot is the concept of zero-moment point, usually abbreviated as ZMP. The concept of ZMP was introduced by Miomir Vukobratović in 1968. It is still the most widely used approach for generating dynamically stable walking movements in which the supporting foot or feet keep contact with the ground surface at all times. This is important to prevent the robot from falling. The basic concepts related to ZMP are described in Sect. 14.1.

Another specific issue that arises when programming humanoid robots is the very high number of degrees of freedom they require compared to standard industrial robots. While typical industrial robots only have 6 and seldom 7 degree of freedom, humanoid robots often have more than 30 degree of freedom. For example, one of the best known humanoid robots Honda Asimo has 34 degree of freedom: 3 in the head, 7 in each arm (3 in the shoulder, 1 in the elbow, and 3 in the wrist), 1 in the waist, 6 in each leg, and 2 in each hand. Such a large number of degrees of freedom makes classical robot programming with teach pendants and textual programming languages impractical. Instead we can exploit the similarity between humanoid robots and humans. Because of this similarity, humanoid robots can perform tasks in a similar way as humans do. This fact gives rise to an idea that instead of programming a humanoid robot, a human teacher can show to the robot how to execute the desired task. The robot can then attempt to replicate the human execution. This way of robot programming is called programming by demonstration or imitation learning. Its successful application requires that a robot transfers the demonstrated motion to its own kinematic and dynamic structure. Furthermore, since natural environments are rarely static but often change, the robot cannot simply replicate the observed movements. Instead, the observed movements should be adapted to the current environmental conditions. These topics are discussed in Sect. 14.2.

14.1 Biped Locomotion

Biped locomotion is an important topic in humanoid robotics. Here we focus on walking, which is distinguished from other forms of biped locomotion such as running by the constraint that at least one foot must always be in contact with the ground. As explained in the introduction, most of the modern humanoid robots exploit the zero-moment point principle to generate stable walking patterns.

14.1.1 Zero-Moment Point

Throughout this section, we assume that the floor is flat and orthogonal to gravity. We start by analyzing the distribution of a vertical component of ground reaction forces (i.e., the component orthogonal to the ground, as shown in Fig. 14.1). The *zero-moment point* is defined as the point where the resultant of these forces intersects with the ground. We first focus on the motion in the sagittal plane (i.e., the plane that divides the body into the left and right part). As shown in Fig. 14.1, a component of the ground reaction force orthogonal to the ground must be positive at all contact points, otherwise the foot would lose contact with the ground as it is not rigidly attached to it. The zero-moment point p_x according to the above definition can be calculated as follows

$$p_x = \frac{\int_{x_b}^{x_f} x f_z(x)\mathrm{d}x}{f_n}, \tag{14.1}$$

$$f_n = \int_{x_b}^{x_f} f_z(x)\mathrm{d}x, \tag{14.2}$$

where $f_z(x)$ is the vertical component of the ground reaction force at contact point x and f_n the net vertical ground reaction force. The reason why p_x is called zero-moment point becomes clear if the moment at p_x is calculated:

$$\begin{aligned}\tau(p_x) &= -\frac{\int_{x_b}^{x_f} (x - p_x) f_z(x)\mathrm{d}x}{f_n} = -\left(\frac{\int_{x_b}^{x_f} x f_z(x)\mathrm{d}x}{f_n} - p_x \frac{\int_{x_b}^{x_f} f_z(x)\mathrm{d}x}{f_n}\right) \\ &= -(p_x - p_x) = 0.\end{aligned} \tag{14.3}$$

Here we integrated the moment $\tau = -(x - p_x) f_z$ across the whole sole area, i.e. $x_b \leq x \leq x_f$. Thus the net moment at the zero-moment point p_x is equal to zero. The zero-moment point is usually abbreviated as ZMP. It is the point on the ground surface where the net angular momentum is equal to zero. If it exists, ZMP is constrained to lie within the support polygon.

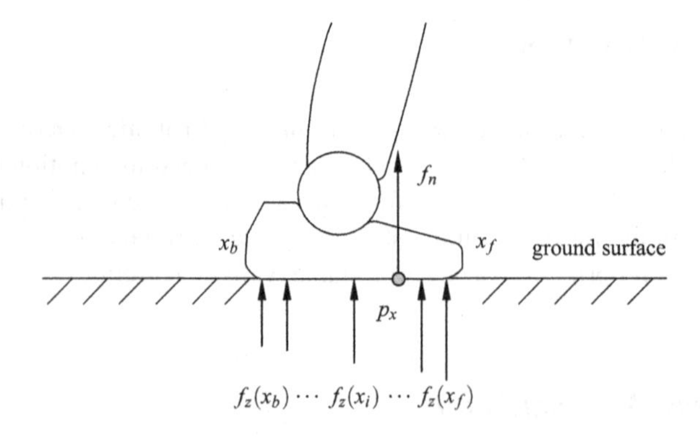

Fig. 14.1 Ground reaction forces $f_z(x_i)$ at different contact points x_i. The zero-moment point p_x and the net ground reaction force orthogonal to the support surface f_n are calculated according to Eqs. (14.1) and (14.2), respectively

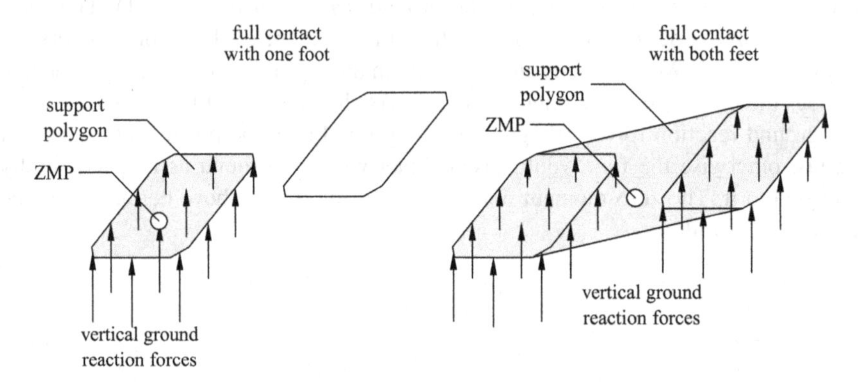

Fig. 14.2 Support polygon (the area enclosed by a gray line) is defined as the convex hull of all points in contact with the ground. Left: support polygon corresponds to the sole area when only one foot is in full contact with the ground. Right: support polygon corresponds to the convex hull of the corners of both feet when both feet are in full contact with the ground

For general humanoid robot walking in 3-D, lateral motion should also be considered. As shown in Fig. 14.2, we must distinguish between two cases: either only one foot is in full contact with the ground or both feet are in full contact with the ground. The ground is assumed to be flat at height p_z. The derivation of ZMP is based on the relationship between the moment about point $\boldsymbol{p} = (p_x, p_y, p_z)$ of the vertical ground reaction force $[0, 0, f_z(\boldsymbol{\xi})]^{\mathrm{T}}$ at all points $\boldsymbol{\xi} = (\xi_x, \xi_y, p_z)$ on the contact surface. The moment is given by

$$\boldsymbol{\tau}(\boldsymbol{p}) = (\boldsymbol{\xi} - \boldsymbol{p}) \times \begin{bmatrix} 0 \\ 0 \\ f_z(\boldsymbol{\xi}) \end{bmatrix} = \begin{bmatrix} (\xi_y - p_y) f_z(\boldsymbol{\xi}) \\ -(\xi_x - p_x) f_z(\boldsymbol{\xi}) \\ 0 \end{bmatrix}. \tag{14.4}$$

To obtain the moment about point $\boldsymbol{p} = (p_x, p_y, p_z)$ due to the orthogonal ground reaction forces $[0, 0, f_z(\boldsymbol{\xi})]^{\mathrm{T}}$ arising at all points of contact $\boldsymbol{\xi}$ between the sole and the ground, we need to integrate across all points of contact

$$\boldsymbol{\tau}_n(\boldsymbol{p}) = \int_S \begin{bmatrix} \xi_x - p_x \\ \xi_y - p_y \\ 0 \end{bmatrix} \times \begin{bmatrix} 0 \\ 0 \\ f_z(\boldsymbol{\xi}) \end{bmatrix} \mathrm{d}S = \begin{bmatrix} \int_S (\xi_y - p_y) f_z(\boldsymbol{\xi}) \mathrm{d}S \\ -\int_S (\xi_x - p_x) f_z(\boldsymbol{\xi}) \mathrm{d}S \\ 0 \end{bmatrix}, \quad (14.5)$$

where S denotes the area of contact. Similarly as in 2D case, the point on the ground where the moment of the normal of the ground reaction force becomes zero (i.e., the zero-moment point $\boldsymbol{\tau}_n(\boldsymbol{p}) = 0$), is given by

$$\boldsymbol{p} = \begin{bmatrix} p_x \\ p_y \\ p_z \end{bmatrix} = \left[\frac{\int_S \xi_x f_z(\boldsymbol{\xi}) \mathrm{d}S}{f_n}, \frac{\int_S \xi_y f_z(\boldsymbol{\xi}) \mathrm{d}S}{f_n}, p_z \right]^{\mathrm{T}}, \quad (14.6)$$

where

$$f_n = \int_S f_z(\boldsymbol{\xi}) \mathrm{d}S \quad (14.7)$$

is the sum of ground reaction forces orthogonal to the ground at all contacts between the sole and the ground.

On a real humanoid robot, ZMP (if it exists), is guaranteed to lie within the support polygon because if the contact between the sole and the ground surface exists, the component of the ground reaction force orthogonal to the ground must be positive. Otherwise the contact between the sole and the ground surface would be lost as the robot is not fixed to the ground and therefore cannot generate negative vertical ground reaction forces. The humanoid robot can control its posture with its feet only if the ZMP exists inside the support polygon. Otherwise the robot loses contact with the ground and cannot control the posture with its feet any more.

14.1.2 Generation of Walking Patterns

In biped walking, the robot's feet alternate between two phases:

- *stance* phase in which the foot's location should not change,
- *swing* phase in which the foot moves.

Figure 14.3 shows these two distinct phases in the gait cycle: when both feet are in contact with the ground, the robot is in *double* support phase. The feet do not move in this phase. Once one of the feet starts moving, the robot transitions from double to *single* support phase, in which one of the two feet moves. The single support phase is followed by another double support phase once the foot in the swing phase establishes a contact with the ground.

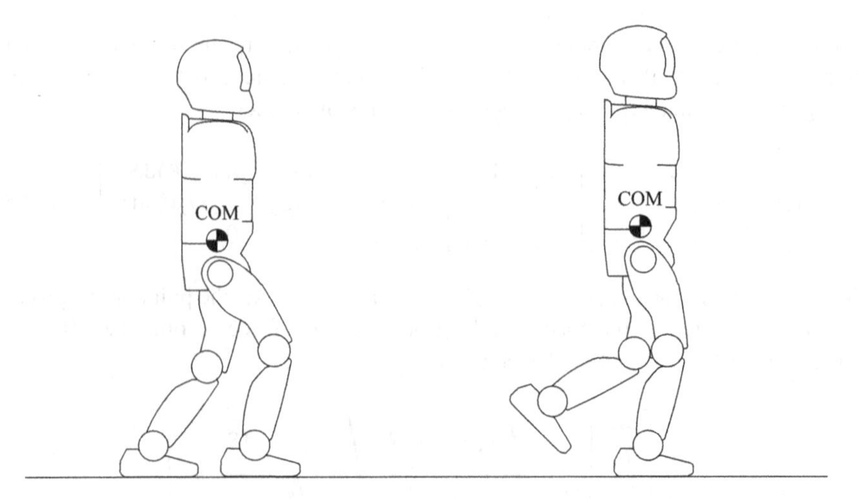

Fig. 14.3 Single and double support phase. In the double support phase, both feet are in contact with the ground and the robot's weight is supported by both legs. In the single support phase, one foot is in motion, whereas the other foot supporting the robot is in contact with the ground

In ZMP-based walking, one or both feet of the robot are always in contact with the ground. Thus ZMP exists and the robot can keep balance by making sure that the support polygon contains the ZMP. However, the robot cannot directly control the ZMP as defined in Eqs. (14.1) and (14.6). We therefore introduce the concept of center of mass (COM). ZMP can be controlled by exploiting its relationship with COM.

Center of mass (COM) is defined as the average position of all body parts of a humanoid robot, weighted with the mass of body parts. For a robot with D rigid links, COM can be calculated as:

$$\boldsymbol{c} = \frac{\sum_{i=1}^{D} m_i \boldsymbol{c}_i}{M}, \quad M = \sum_{i=1}^{D} m_i, \tag{14.8}$$

where m_i is the mass of i-th link and $\boldsymbol{c}_i$ its position, which can be calculated by direct kinematics provided the center of mass of each link is known in the link's local coordinates. With some approximations, the relationship between ZMP and COM can be specified as follows

$$p_x = c_x - \frac{(c_z - p_z)\ddot{c}_x}{\ddot{c}_z + g}, \tag{14.9}$$

$$p_y = c_y - \frac{(c_z - p_z)\ddot{c}_y}{\ddot{c}_z + g}, \tag{14.10}$$

where p_z denotes the height of the ground floor, g is the gravity constant, and $\boldsymbol{c} = (c_x, c_y, c_z)$ and $\boldsymbol{p} = (p_x, p_y, p_z)$ are the coordinates of COM and ZMP, respectively. Note that if the robot is at rest, i.e. $\ddot{c}_x = \ddot{c}_y = 0$, then ZMP and the projection of COM coincide as $p_x = c_x$ and $p_y = c_y$. Note also that if the ground is flat and orthogonal to gravity, as we assumed in Sect. 14.1.1, p_z is a constant.

In general we distinguish between *static* and *dynamic walking*. Static walking is defined as any stable walking motion where the projection of COM always stays inside the support polygon. This means that if the robot completely stops moving at any moment during walking, it does not fall down because for the robot at rest, the projection of COM onto the ground surface is equal to the ZMP (see Eqs. (14.9) and (14.10)). In static walking the motion must generally be slow so that the projection of COM is close to the ZMP. This kind of walking typically requires large feet and strong ankle joints to generate sufficient forces at the ankles. As the robot's motion becomes faster, ZMP and the projection of COM become more different and stability cannot be ensured by controlling the projection of COM only.

More effective walking behaviors are generated by dynamic walking patterns, where the projection of COM is not equal to ZMP and can fall outside of the support polygon during some period of motion. A ZMP-based dynamic walking pattern is shown in Fig. 14.4. Such patterns are planned so that the ZMP remains within the boundary of the support polygon in all phases of walking. This can be accomplished as follows:

- Specify the Cartesian motion of the robot's feet. Here the robot's step length and timing of foot motion is prescribed.
- Specify the reference ZMP trajectory so that ZMP remains within the support polygon at all times.
- Determine the humanoid robot's upper body motion in order to realize the reference ZMP motion. This can be accomplished using Eqs. (14.9) and (14.10).
- The humanoid robot's leg motion is finally calculated from the body and feet motion using inverse kinematics.

The motion of COM is not fully specified by Eqs. (14.9) and (14.10) as there are only two equations and three unknown parameters. To fully specify the motion of COM and consequently the motion of the humanoid robot's upper body, an additional constraint must be imposed. There are several possible approaches. The simplest among them is to set the height of COM to a constant value (i.e., $c_z = \text{const}, \ddot{c}_z = 0$). With this assumption, the motion of COM is fully specified by Eqs. (14.9) and (14.10). A more adaptable and active motion can be achieved if c_z is allowed to vary.

Note that the above approach determines the motion of COM without considering the legs. However, since most of the mass is usually concentrated in the upper body of a humanoid robot and since it is not necessary to follow the prescribed ZMP trajectory exactly, the above approach is sufficient to generate dynamically stable walking patterns.

If an accurate model of the robot is available, biped walking can be realized by simply following a predetermined walking pattern. Due to noise and model inaccuracies, in practice such an approach usually does not result in a stable walking behavior

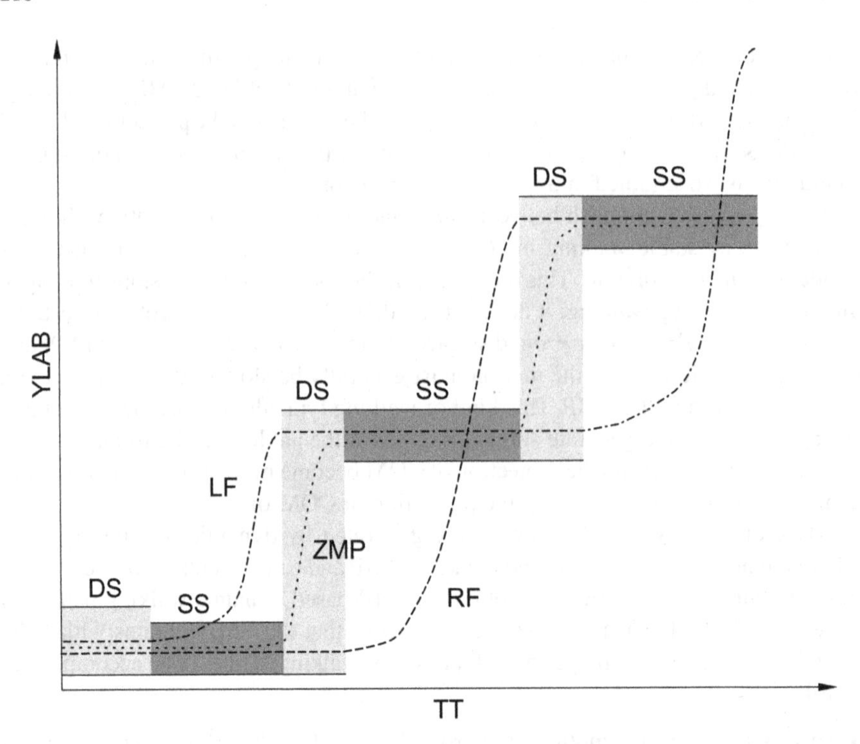

Fig. 14.4 An example ZMP-based walking pattern in sagittal plane. The robot starts with both feet placed roughly parallel on the ground and then generates three steps, starting with the left foot. The shaded areas show the extent of support polygon during single support phase (dark shaded area) and double support phase (light shaded area). The ZMP trajectory (dotted) is planned in such a way that it remains within the support polygon during the whole duration of walking. The trajectories of both feet (left: dashed dotted, right: dashed) are also shown

without supplementing the precomputed walking pattern with a stabilizer that modifies the pattern according to the sensory input provided by gyros, accelerometers, force sensors, cameras, etc.

It should be pointed out that ZMP is not the only principle that can be used to generate stable walking patterns. It is possible to generate a walking pattern where a robot is unstable for some period of motion. Such walking patterns must be planned so that the robot can recover from instabilities before falling to the ground.

14.2 Imitation Learning

To fully exploit their potential, humanoid robots should be able to perform a variety of tasks in unstructured environments (e.g., people's homes, hospitals, shops, offices, and even outdoor environments). The aforementioned robotics challenge was geared towards humanoid robots at disaster sites. Unlike many industrial environments, where robots are widely used today, such environments cannot be prepared in advance to ease the operation of a humanoid robot. The programming of humanoid robots is further complicated by the large number of degrees of freedom involved in humanoid robot motion. Hence classic robot programming techniques based on teach pendants, carefully prepared off-line simulation systems, and programming languages are not sufficient for humanoids. Instead, it is necessary to equip humanoid robots with learning and adaptation capabilities. This way they can be programmed more easily and even autonomously acquire additional knowledge.

Learning of humanoid robot behaviors is a difficult problem because the space of all humanoid robot motions that needs to be explored is very large and increases exponentially with the number of degrees of freedom. A solution to this problem is to focus learning on those parts of the robot motion space that are actually relevant for the desired task. This can be achieved by *imitation learning*, also referred to as *programming by demonstration*. With this approach, a human teacher demonstrates to a robot how to perform the desired task. For it to work, a robot must be able to extract the important information from human demonstration and replicate the essential parts of task execution. While in most cases it is not necessary to exactly replicate the demonstrated movements to successfully execute the desired task, it is advantageous if the robot can mimic the demonstrated movement as much as possible. Since the body of a humanoid robot is similar to a human body, imitation learning is often a good approach to focus learning on the relevant parts of humanoid robot motion space.

14.2.1 Observation of Human Motion and Its Transfer to Humanoid Robot Motion

There are many possible measurement systems and technologies that can be used to observe and measure human movements. They include

- optical motion capture systems,
- ensembles of inertial measurement units (IMU),
- computer vision methods for the estimation of human motion,
- passive exoskeletons,
- hand guiding.

In the following we explain the major advantages and drawbacks of these systems.

14.2.1.1 Optical Tracking Devices for Human Motion Capture

Optical trackers are based on a set of markers attached to a human body. Markers can either be passive or active. Passive markers are made of retroreflective materials, which reflect light in the direction from where it came. In systems with passive markers, cameras are equipped with a band of infrared light emitting diodes (LEDs). The emitted light bounces off the marker back in the direction of the camera, making the marker much brighter than any other point in the image. This property makes retroreflective markers easy to detect in camera images. Using triangulation, a 3D marker location can be calculated if the marker is detected in at least two simultaneously acquired camera images. The predicted motion of visible markers is used to match the visible markers extracted at two successive measurement times.

Unlike passive markers that reflect light, active markers are equipped with LEDs and thus emit their own light. Consequently they must be powered. Optical trackers with active markers usually illuminate only one marker at a time for a very short time. Thus the system always knows which marker is currently visible, thereby providing the identity of the marker. For this reason, optical tracking systems with active markers can cope with temporary occlusions more effectively than systems with passive markers because an occluded active marker can be identified once it becomes visible again. This is not the case with passive markers. On the other hand, since active markers require power, they need to be connected to a power source with cables. This makes them more cumbersome to use than passive markers that require no cables.

To measure human motion, both passive and active markers must be attached to the human body segments at appropriate locations. Usually at least three markers are attached to each body segment, otherwise the location of rigid body segments cannot be estimated. Various special motion capture suits were designed in the past to ease the attachment of markers to the relevant body segments.

Optical tracking systems with active or passive markers provide 3-D locations of markers attached to the human body that are currently in view. The 3-D position and orientation of a body segment can be estimated if at least three markers attached to the segment are visible. In order to reproduce the observed motion with a robot, this information needs to be related to the robot motion. To a certain degree of accuracy, human motion can be modelled as an articulated motion of rigid body parts. If a humanoid robot kinematics is close enough to the human body kinematics, we can embed it into a human body as shown in Fig. 14.5. Such an embedding can later be used to estimate the joint angles from the orientations of successive body segments. Let us assume that the orientation of two successive body segments is given by orientation matrices $\mathbf{R}_1$ and $\mathbf{R}_2$ and that the joint linking the two segments consists of three successive joint axes $\mathbf{j}_1$, $\mathbf{j}_2$ and $\mathbf{j}_3$ with rotation angles denoted by φ, θ, and ψ, respectively. We further assume that two consecutive joint axes are orthogonal and that all three axes intersect in a common point. In such an arrangement, the three joint angles correspond to the Euler angles introduced in Chap. 4. There are altogether 12 different joint axis combinations that cover every possible arrangement of axes in joints with three degrees of freedom. In Fig. 14.5, torso, neck, shoulder, wrist, and ankle joints can be described by an appropriate combination of Euler angles. The relationship between these values is given by

$$\mathbf{R}_1 = \mathbf{R}(\mathbf{j}_1, \varphi)\mathbf{R}(\mathbf{j}_2, \theta)\mathbf{R}(\mathbf{j}_3, \psi)\mathbf{R}_2 = \mathbf{R}(\varphi, \theta, \psi)\mathbf{R}_2. \tag{14.11}$$

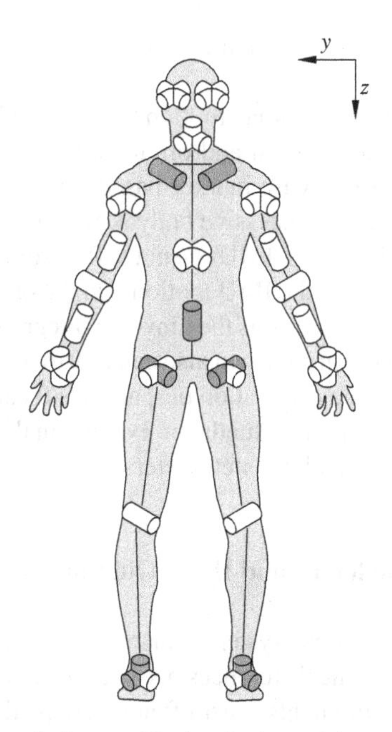

Fig. 14.5 Kinematic structure of a humanoid robot. In the upright position with extended arms and legs, all joint axes are parallel to one of the three main axes of the body (forward/backward: x axis, left/right: y axis, up/down: z axis)

The joint angles ϕ, θ, and ψ can then be calculated by solving equation

$$\mathbf{R}(\varphi, \theta, \psi) = \mathbf{R}_1\mathbf{R}_2^{\mathrm{T}}. \tag{14.12}$$

This equation depends on the choice of joint axes $\mathbf{j}_1$, $\mathbf{j}_2$, and $\mathbf{j}_3$. The observed motion can be replicated by a robot once all relevant joint angles from the embedded model have been estimated.

Optical tracking systems can also accurately estimate the absolute position and orientation of the human body in a world coordinate system. As the root of a humanoid robot's kinematics is typically assumed to be at the local coordinate frame attached to the torso, the estimated position and orientation of the torso corresponds to the absolute position and orientation of the human body in world coordinates.

14.2.1.2 Inertial Measurement Units (IMUs)

Inertial measurement units (IMUs) contain different sensors including accelerometers to measure 3D linear acceleration and gyroscopes for measuring the rate of change of 3D orientation (i.e., angular velocity). IMUs also often include magnetometers to provide redundant measurements to improve accuracy and reduce the drift. From these data, the position and orientation of an IMU can be estimated as explained in Sect. 7.2.6.

In the context of transferring human motion to humanoid robot motion, IMU data can be used to estimate the position and orientation of each body segment which has an IMU attached. Just like with marker-based trackers, the joint angles can be estimated from orientations of successive body segments using Eq. (14.12).

Unlike optical tracking systems, IMUs do not suffer from occlusions as no external cameras are needed to measure the IMU motion. On the other hand, IMUs are not as accurate as optical tracking systems as they involve integration of linear acceleration and angular velocity. The integration can also cause drift, especially when estimating the absolute body position and orientation in space. Drift can be reduced by developing appropriate filters that exploit redundancy existing in the measurements obtained from accelerometers, gyros, and magnetometers.

14.2.1.3 Passive Exoskeletons and Hand Guiding

A crucial issue that all of the above systems must deal with is that they measure human motion without considering the differences between the human and robot kinematics and dynamics. Such measurements must often be adapted to the robot constraints, otherwise the robot cannot execute the demonstrated movements. Alternatively, a nonlinear optimization problem can be formulated to adapt the demonstrated motion to the capabilities of a target robot.

The problem of transferring human motion to robot motion can be avoided by applying different measurement systems. One possibility is to design a special passive device, which is worn like exoskeleton with the degrees of freedom that correspond to the robot degrees of freedom. The passive exoskeleton must be designed in such a way that it does not restrict motion for most movements. It has no motors, but it should be equipped with goniometers to measure the joint angles. The joint angles measured by the exoskeleton can be used to directly control the robot if the kinematics of the target robot corresponds to the kinematic of the exoskeleton. One drawback of passive exoskeletons is that like clothes, they must be built to the specific size of a human demonstrator.

As explained in Sect. 12.3.2, some robots can be physically guided through the desired movements (see also Fig. 14.6). During hand guiding the movement is recorded by the robot's own joint angle sensors and is thus by default kinematically feasible. This approach is effective if the robot is compliant and can compensate for gravity, so that a human demonstrator can easily move it in the desired direction.

The main drawback of hand guiding is that the demonstration of the desired motion is less natural for a human demonstrator than for example when marker-based tracking systems are used. Thus with such systems it is sometimes not as easy to demonstrate complex movements. For example, hand guiding is not effective to demonstrate complex dancing movements. On the other hand, dancing can be easily demonstrated by a human directly and measured with an optical tracker, IMUs, or a passive exoskeleton.

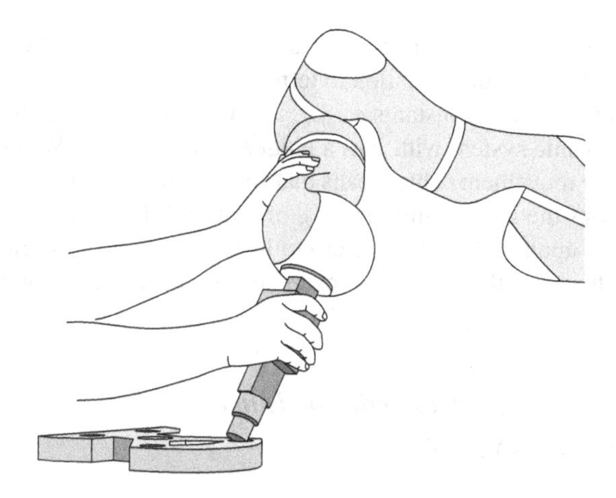

Fig. 14.6 Demonstration of peg-in-hole task by kinesthetic teaching. Human demonstrator guides the anthropomorphic arm through the task execution with its own hands

14.2.2 Dynamic Movement Primitives

In Sect. 14.2.1 we discussed how to measure human demonstrations and how to transform the measured movements into the robot joint angle trajectories. In some cases it is also necessary to adapt the measured motion to the kinematic and dynamic capabilities of the target robot. Typically, we end up with a measurement sequence

$$\{\mathbf{y}_d(t_j), t_j\}_{j=1}^T, \tag{14.13}$$

where $\mathbf{y}_d(t_j) \in \mathbb{R}^D$ are the measured joint angles at time t_j, D is the number of degrees of freedom, and T is the number of measurements. This sequence defines the reference trajectory. However, for effective control we need to generate motor commands with the servo rate of the target robot. The robot's servo rate is often higher than the capture rate of the measurement system. Thus from the measurement data (14.13) we need to generate a continuous reference trajectory in order to generate motor commands to control the robot at the appropriate rate.

In this section we introduce *Dynamic Movement Primitives* (DMPs), which provide a comprehensive framework for the effective imitation learning and control of robot movements. DMPs are based on a set of nonlinear differential equations with well-defined attractor dynamics. For a single robot degree of freedom, here denoted by y and taken to be one of the D recorded joint angles, the following system of linear differential equations with constant coefficients is analyzed to derive a DMP

$$\tau \dot{z} = \alpha_z(\beta_z(g - y) - z), \tag{14.14}$$

$$\tau \dot{y} = z. \tag{14.15}$$

Note that auxiliary variable z is just a scaled velocity of the control variable y. Constants α_z and β_z have an interpretation in terms of spring stiffness and damping. For the appropriately selected constants $\alpha_z, \beta_z, \tau > 0$, these equations form a globally stable linear dynamic system with g as a unique point attractor. We often refer to g as the *goal* of the movement. This means that for any start configuration $y(0) = y0$, variable y reaches the goal configuration g after a certain amount of time, just as a stretched spring, upon release, will return to its rest position. τ is referred to as the *time constant*. It affects the speed of convergence to the attractor point g.

14.2.3 Convergence Properties of Linear Dynamic Systems

Let us analyze why the above system is useful. We start by writing down a general solution of the non-homogenous linear differential equation system (14.14) and (14.15). It is well known that the general solution of such a system can be written as a sum of the particular and homogeneous solution

$$\begin{bmatrix} z(t) \\ y(t) \end{bmatrix} = \begin{bmatrix} z_p(t) \\ y_p(t)] \end{bmatrix} + \begin{bmatrix} z_h(t) \\ y_h(t) \end{bmatrix}. \tag{14.16}$$

Here $[z_p(t), y_p(t)]^{\mathrm{T}}$ denotes any function that solves the linear system (14.14)–(14.15), while $[z_h(t), y_h(t)]^{\mathrm{T}}$ is the general solution of the homogeneous part of Eqs. (14.14)–(14.15), i.e.,

$$\begin{bmatrix} \dot{z} \\ \dot{y} \end{bmatrix} = \frac{1}{\tau} \begin{bmatrix} -\alpha_z(\beta_z y + z) \\ z \end{bmatrix} = \mathbf{A} \begin{bmatrix} z \\ y \end{bmatrix}, \quad \mathbf{A} = \frac{1}{\tau} \begin{bmatrix} -\alpha_z & -\alpha_z\beta_z \\ 1 & 0 \end{bmatrix}.$$

It is easy to check that constant function $[z_p(t), y_p(t)]^{\mathrm{T}} = [0, g]^{\mathrm{T}}$ solves the equation system (14.14) and (14.15). Additionally, it is well known that the general solution of homogeneous system (14.17) is given by $[z_h(t), y_h(t)]^{\mathrm{T}} = \exp(\mathbf{A}t)\,\mathbf{c}$, where $\mathbf{c} \in \mathbb{R}^2$ is an arbitrary constant. Thus, the general solution of Eqs. (14.14) and (14.15) can be written as

$$\begin{bmatrix} z(t) \\ y(t) \end{bmatrix} = \begin{bmatrix} 0 \\ g \end{bmatrix} + \exp(\mathbf{A}t)\,\mathbf{c}. \tag{14.17}$$

Constant $\mathbf{c}$ should be calculated from the initial conditions, $[z(0), y(0)]^{\mathrm{T}} = [z0, y0]^{\mathrm{T}}$. The eigenvalues of $\mathbf{A}$ are given by $\lambda_{1,2} = \left(-\alpha_z \pm \sqrt{\alpha_z^2 - 4\alpha_z\beta_z}\right)/(2\tau)$. Solution (14.17) converges to $[0, g]^{\mathrm{T}}$ if the real part of eigenvalues $\lambda_{1,2}$ is smaller than 0, which is true for any $\alpha_z, \beta_z, \tau > 0$. The system is critically damped, which means that y converges to g without oscillating and faster than for any other choice of $\mathbf{A}$, if $\mathbf{A}$ has two equal negative eigenvalues. This happens at $\alpha_z = 4\beta_z$ where $\lambda_{1,2} = -\alpha_z/(2\tau)$.

14.2.4 Dynamic Movement Primitives for Point-to-Point Movements

Differential equation system (14.14)–(14.15) ensures that y converges to g from any starting point y_0. It can therefore be used to realize simple point-to-point movements. To increase a rather limited set of trajectories that can be generated by (14.14) and (14.15) and thus enable the generation of general point-to-point movements, we can add a nonlinear component to Eq. (14.14). This nonlinear function is often referred to as *forcing term*. A standard choice is to add a linear combination of radial basis functions Ψ_i

$$f(x) = \frac{\sum_{i=1}^{N} w_i \Psi_i(x)}{\sum_{i=1}^{N} \Psi_i(x)} x(g - y_0), \tag{14.18}$$

$$\Psi_i(x) = \exp\left(-h_i (x - c_i)^2\right), \tag{14.19}$$

where c_i are the centers of radial basis functions distributed along the phase of the trajectory and $h_i > 0$. The term $g - y_0$, $y_0 = y(t_1)$, is used to scale the trajectory if the initial and / or final configuration change. As long as the beginning and the end of movement are kept constant, this scaling factor has no effect and can be omitted. *Phase* variable x is used in forcing term (14.18) instead of time to make the dependency of the resulting control policy on time more implicit. Its dynamics is defined by

$$\tau \dot{x} = -\alpha_x x, \tag{14.20}$$

with the initial value $x(0) = 1$. A solution to (14.20) is given by

$$x(t) = \exp\left(-\alpha_x t / \tau\right). \tag{14.21}$$

The appealing property of using the phase variable x instead of explicit time is that by appropriately modifying Eq. (14.20), the evolution of time can be stopped to account for perturbations during motion. There is no need to manage the internal clock of the system. We obtain the following system of nonlinear differential equations

$$\tau \dot{z} = \alpha_z(\beta_z(g - y) - z) + f(x), \tag{14.22}$$

$$\tau \dot{y} = z. \tag{14.23}$$

The phase variable x and consequently $f(x)$ tend to 0 as time increases. Hence the influence of nonlinear term $f(x)$ decreases with time. Consequently, through the integration of system (14.22)–(14.23) the system variables $[z, y]^T$ are guaranteed to converge to $[0, g]^T$, just like the linear system (14.14)–(14.15). The control policy specified by variable y and its first- and second-order derivatives defines what we call a *dynamic movement primitive* (DMP). For a system with many degrees of freedom, each degree of freedom is represented by its own differential equation system (14.22)–(14.23), whereas the phase x is common across all the degrees of freedom. This can be done because phase Eq. (14.20) does not include variables y and z.

It is usually sufficient to determine the parameters c_i and h_i of Eq. (14.19) by setting a predefined distribution pattern and increasing the number of base functions N until the desired reconstruction accuracy can be achieved. For example, for a given N we can define

$$c_i = \exp\left(-\alpha_x \frac{i-1}{N-1}\right), \; i = 1, \ldots, N, \tag{14.24}$$

$$h_i = \frac{2}{(c_{i+1} - c_i)^2}, \; i = 1, \ldots, N-1, \; h_N = h_{N-1}. \tag{14.25}$$

Note that $c_1 = 1 = x(0)$ and $c_N = \exp(-\alpha_x) = x(t_T)$.

In the equations above, α_x, α_z, and β_z are set to constant values. The values must be chosen in such a way that the convergence of the underlying dynamic system is ensured as explained in Sect. 14.2.3. This is the case if we set $\alpha_x = 2$, $\beta_z = 3$, $\alpha_z = 4\beta_z = 12$.

DMPs were designed to provide a representation that enables accurate encoding of the desired point-to-point movements and at the same time permits modulation of different properties of the encoded trajectory. In this context, the shape parameters w_i are determined so that the robot can accurately follow the desired trajectory by integrating the equation system (14.20), (14.22), and (14.23). The other parameters are used for modulation and to account for disturbances.

For a movement with two degrees of freedom, Fig. 14.7 shows a graphical plot of attractor fields generated by the dynamic movement primitive. The attractor field changes with the evolution of phase x. As long as the robot follows the demonstrated trajectory, the attractor field directs the robot to move along the demonstrated trajectory. However, if the robot is perturbed and deviates from the demonstrated trajectory, the attractor fields generated along the phase x directs the robot so that it reaches the desired final configuration (goal), albeit along a modified trajectory.

A trajectory can be reproduced from a fully specified DMP by integrating Eqs. (14.22), (14.23), and (14.20) using Euler integration method:

$$z_{k+1} = z_k + \frac{1}{\tau}\left(\alpha_z(\beta_z(g - y_k) - z_k) + f(x_k)\right)\Delta t, \tag{14.26}$$

$$y_{k+1} = y_k + \frac{1}{\tau} z_i \Delta t, \tag{14.27}$$

$$x_{k+1} = x_k - \frac{1}{\tau}\alpha_x x_k \Delta t, \tag{14.28}$$

where $\Delta t > 0$ is the integration constant usually set to the robot's servo rate. The initial parameters for integration must be set to the current state of the robot, which

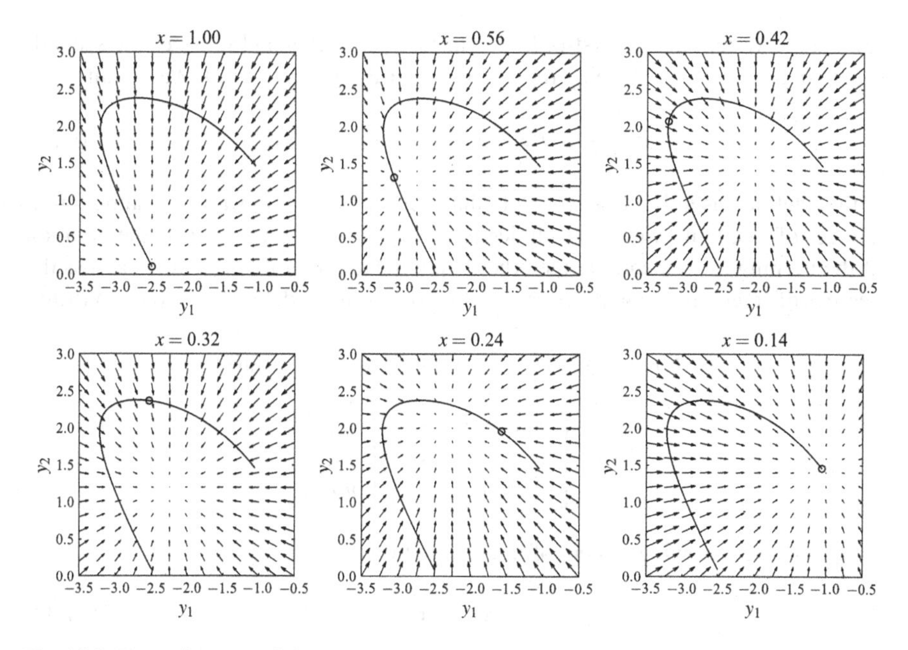

Fig. 14.7 Plots of attractor fields generated by a DMP specifying the motion of a robot with two degrees of freedom y_1, y_2 as it is integrated along the phase x. The arrows in each plot show $\dot{z}_1$, $\dot{z}_2$ at different values of y_1, y_2 at the given phase x, assuming that only y_1 and y_2 have changed compared to the unperturbed trajectory. The circles show the desired configurations y_1, y_2 at the given phase x

at the beginning of motion is assumed to be at the given initial position and with zero velocity. This results in the following initialization formulas: $y_0 = y0,\ z_0 = 0,\ x = 1$.

14.2.5 Estimation of DMP Parameters from a Single Demonstration

To estimate the DMP representing the measurement sequence (14.13), we first compute the derivatives $\dot{\mathbf{y}}_j$ and $\ddot{\mathbf{y}}_j$ by numerical differentiation. For any of the D degrees of freedom y, we obtain the following measurement sequence

$$\{y_\mathrm{d}(t_j),\ \dot{y}_\mathrm{d}(t_j),\ \ddot{y}_\mathrm{d}(t_j)\}_{j=1}^{T}, \tag{14.29}$$

where $y_d(t_j),\ \dot{y}_d(t_j),\ \ddot{y}_d(t_j) \in \mathbb{R}$ are the measured positions, velocities, and accelerations on the training trajectory and T is the number of sampling points. Using the DMP movement representation, the trajectory of any smooth movement can be approximated by estimating parameters w_i of Eq. (14.18). For this purpose we rewrite the system of two first-order linear Eqs. (14.22) and (14.23) as one second-order equation. This is done by replacing z with $\tau \dot{y}$ in Eq. (14.22). We obtain

$$\tau^2 \ddot{y} + \alpha_z \tau \dot{y} - \alpha_z \beta_z (g - y) = f(x), \tag{14.30}$$

with f defined as in Eq. (14.18). Note that time constant τ must be the same for all degrees of freedom. A possible choice is $\tau = t_T - t_1$, where $t_T - t_1$ is the duration of the training movement. On the other hand, the attractor point g varies across the degrees of freedom. It can be extracted directly from the data: $g = y_d(t_T)$. Writing

$$F_d(t_j) = \tau^2 \ddot{y}_d(t_j) + \alpha_z \tau \dot{y}_d(t_j) - \alpha_z \beta_z (g - y_d(t_j)), \tag{14.31}$$

$$\mathbf{f} = \begin{bmatrix} F_d(t_1) \\ \dots \\ F_d(t_T) \end{bmatrix}, \quad \mathbf{w} = \begin{bmatrix} w_1 \\ \dots \\ w_N \end{bmatrix},$$

we obtain the following system of linear equations

$$\mathbf{X}\mathbf{w} = \mathbf{f}, \tag{14.32}$$

which must be solved to estimate the weights of a DMP encoding the desired motion. The system matrix $\mathbf{X}$ is given by

$$\mathbf{X} = (g - y_0) \begin{bmatrix} \frac{\Psi_1(x_1)}{\sum_{i=1}^N \Psi_i(x_1)} x_1 & \dots & \frac{\Psi_N(x_1)}{\sum_{i=1}^N \Psi_i(x_1)} x_1 \\ \dots & \dots & \dots \\ \frac{\Psi_1(x_T)}{\sum_{i=1}^N \Psi_i(x_T)} x_T & \dots & \frac{\Psi_N(x_T)}{\sum_{i=1}^N \Psi_i(x_T)} x_T \end{bmatrix}. \tag{14.33}$$

The phase sampling points x_j are obtained by inserting measurement times t_j into Eq. (14.21). The parameters $\mathbf{w}$ can be calculated by solving the above system of linear equations in a least-squares sense. An example DMP estimation is shown in Fig. 14.8. The calculated DMP ensures that the robot reaches the attractor point g at time t_T. Since DMPs have been designed to represent point-to-point movements, the demonstrated movement must come to a full stop at the end of the demonstration if the robot is to stay at the attractor point after t_T. If any other type of motion is approximated by a DMP, the robot will overshoot the attractor point and return back to it after the dynamics of the second-order linear system of differential equations starts dominating the motion. At least theoretically, the velocity does not need to be zero at the beginning of movement, but it is difficult to imagine a real programming by demonstration system in which such a trajectory would be acquired.

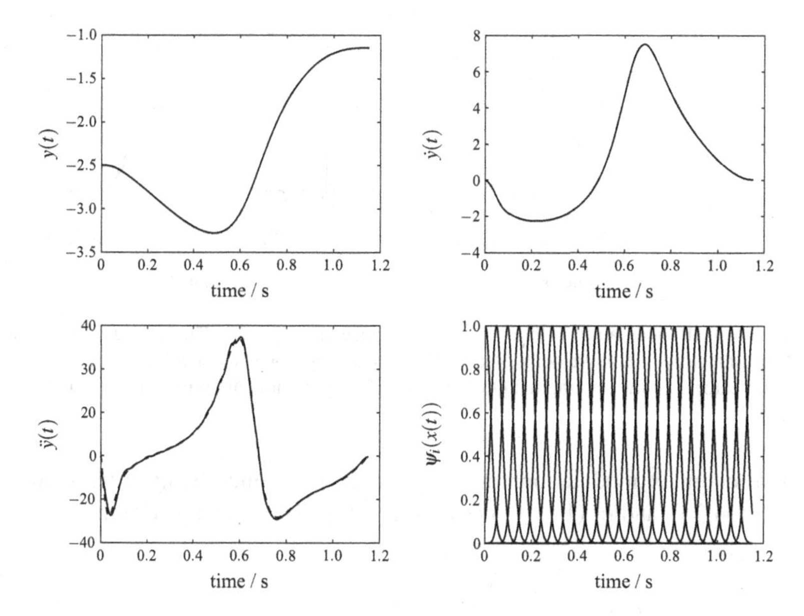

Fig. 14.8 Time evolution of an example dynamic movement primitive: control variable y and its derivatives, phase x, and radial basis functions ψ_i are all shown with solid lines. Dashed lines show the demonstrated values of y, $\dot{y}$ and $\ddot{y}$

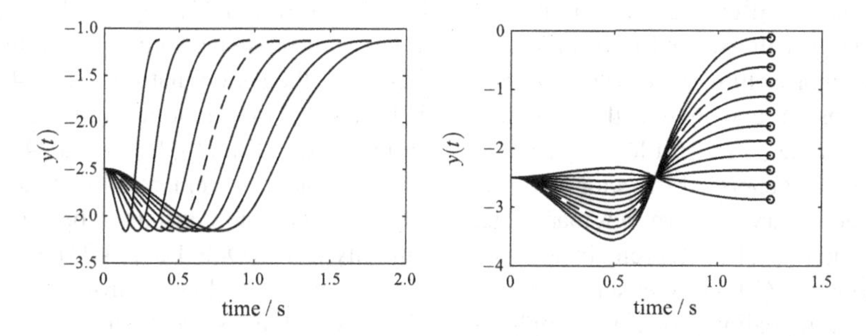

Fig. 14.9 DMP modulations. The dashed trajectories show the original DMP without applying any modulation. **Left**: Time modulation. Solid trajectories show DMPs with changed τ. **Right**: Goal modulation. Solid trajectories show DMPs with changed goal g. Circles show the goal position

14.2.6 Modulation of DMPs

An important advantage of DMPs is that they enable easy modulation of the learnt movement. Figure 14.9 left shows that by changing parameter τ, the movement can be sped up or slowed down. The same figure also shows that by changing the goal parameter g, the final configuration on the trajectory can be changed so that the robot moves to a new goal. The term $y_0 - g$ in the forcing term (14.18) ensures that the movement is appropriately scaled as the goal or initial configuration changes.

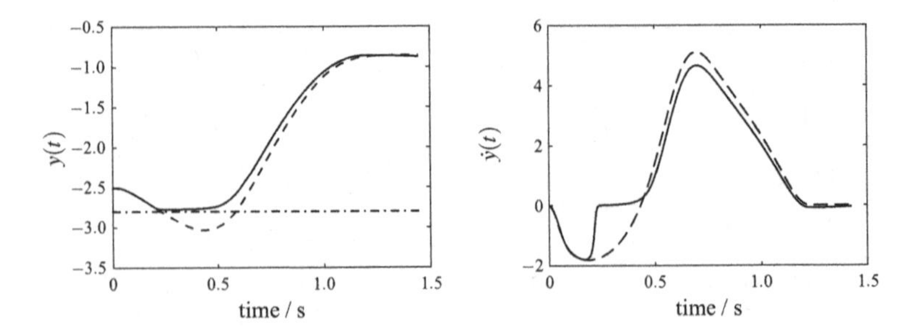

Fig. 14.10 DMP modulation with joint limit avoidance at $y = -2.8$. The solid trajectories show the DMP trajectory and its velocity obtained by integrating (14.34) instead of (14.23), while the dashed trajectories show the original DMP and its velocity without applying any modulation

More complex modulations involve changing the underlying differential Eqs. (14.22), (14.23), and/or (14.20). For example, Eq. (14.23) can be changed to

$$\tau y = z - \frac{\rho}{(y_L - y)^3} \tag{14.34}$$

to implement the avoidance of a lower joint limit. This happens because once y starts approaching y_L, the denominator in Eq. (14.34) becomes small and there is a significant difference between integrating Eq. (14.23) or (14.34). Figure 14.10 right shows that the second term in Eq. (14.34) acts as a repulsive force, preventing y from approaching y_L too closely. On the other hand, the denominator in Eq. (14.34) remains large as long as the joint angle y is far away from the joint limit y_L. Thus in this case there is little difference between integrating Eq. (14.23) or (14.34) and the DMP generated trajectory follows the demonstrated movement. Note that it is not necessary to learn new parameters w_i, goal g, or time constant τ because of modulation. They can remain as they were initially learnt. Only Eq. (14.23) must be changed to (14.34) to ensure joint limit avoidance during on-line control.

The appealing property of applying the phase variable instead of time is that we can easily modulate the time evolution of phase, e.g., by speeding up or slowing down a movement as appropriate by means of coupling terms. Instead of integrating Eqs. (14.20) and (14.23) at time of execution, the modified Eqs. (14.20) and (14.36) could be integrated

$$\tau \dot{x} = -\frac{\alpha_x x}{1 + \alpha_{px}(y - \tilde{y})^2}, \tag{14.35}$$

$$\tau \dot{y} = z + \alpha_{py}(y - \tilde{y}), \tag{14.36}$$

where y and $\tilde{y}$ respectively denote the desired and actual robot joint angle position, respectively. If the robot cannot follow the desired motion, $\alpha_{px}(y - \tilde{y})^2$ becomes large, which in turn makes the phase change $\dot{x}$ small. Thus the phase evolution is stopped until the robot catches up with the desired configuration y. This will

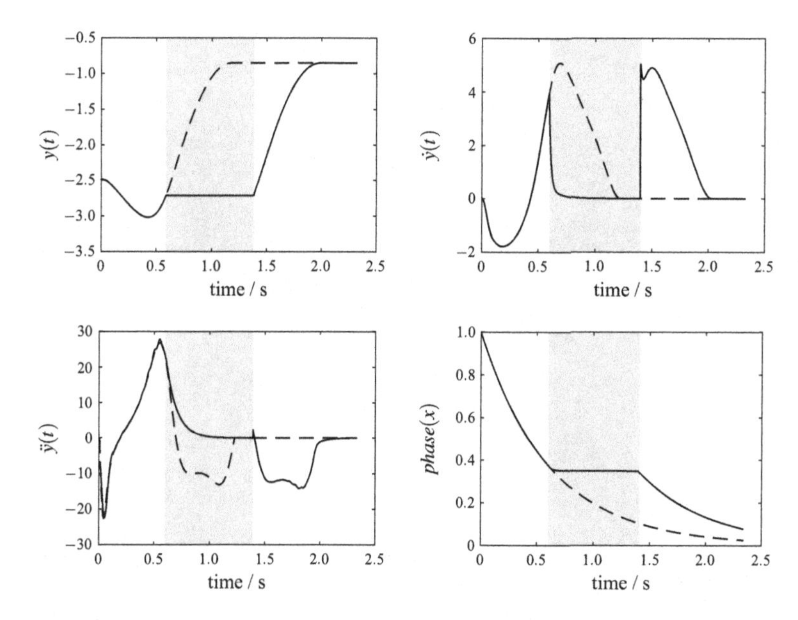

Fig. 14.11 The effect of phase stopping caused by blocking the evolution of joint position $\tilde{y}$ in the time interval [0.6, 1.4] (grey area). The dashed trajectories show the original DMP, velocity, acceleration and phase evolution, while the solid trajectories show their counterparts from the perturbed motion with phase stopping enabled. Note that outside of the time interval [0.6, 1.4] where the joint motion is blocked, the robot accurately follows the desired motion

eventually happen due to the added term in Eq. (14.36). On the other hand, if the robot follows the desired movement precisely, then $\tilde{y} - y \approx 0$ and Eqs. (14.35) and (14.36) are no different from Eqs. (14.20) and (14.23), respectively. Thus in this case the DMP-generated movement is not altered. Figure 14.11 illustrates the effect of phase stopping when the robot's motion is temporarily blocked.

In summary, DMPs provide an effective representation for learning humanoid robot trajectories and to control humanoid robots. They are based on autonomous, nonlinear differential equations that are guaranteed to create smooth kinematic control policies. An important property of DMPs is that they can be learnt from a single demonstration of the desired task. They have several advantages compared to other motor representations including

- they possess free parameters that are easy to learn in order to reproduce any desired movement,
- they are not explicitly dependent on time and allow for time modulation,
- they are robust against perturbations,
- they are easy to modulate by adapting various parameters and equations.

Due to their flexibility and robustness, DMPs are considered a method of choice when learning robot trajectories from single demonstrations.

Chapter 15
Accuracy and Repeatability of Industrial Manipulators

In this chapter we shall briefly consider performance criteria and the methods for testing of industrial robot manipulators as described in the ISO 9283 standard. Before addressing accuracy and repeatability of industrial manipulators we will summarize basic information about robot manipulators.

The basic robot data typically includes a schematic drawing of the robot mechanical structure:

- cartesian robot (Fig. 15.1 left),
- cylindrical robot (Fig. 15.1 right),
- polar (spherical) robot (Fig. 15.2 left),
- anthropomorphic robot (Fig. 15.2 right),
- SCARA robot (Fig. 15.3).

In all drawings the degrees of freedom of the robot mechanism must be marked. The drawing must include also the base coordinate frame and the mechanical interface frame which are determined by the manufacturer.

Of special importance is the diagram showing the boundaries of the workspace (Fig. 15.4). The maximal reach of the robot arm must be clearly shown in at least two planes. The range of motion for each robot axis (degree of freedom) must be indicated. The manufacturer must specify also the center of the workspace c_w, where most of the robot activities take place.

The robot data must be accompanied by the characteristic loading parameters, such as mass (kg), torque (Nm), moment of inertia (kgm^2), and thrust (N). The maximal velocity must be given at a constant rate, when there is no acceleration or deceleration. The maximal velocities for particular robot axes must be given with the load applied to the end-effector. The resolution of each axis movement (mm or °), description of the control system and the programming methods must also be presented.

The three most relevant robot coordinate frames (right-handed) are shown in Fig. 15.5. First is the world coordinate frame x_0–y_0–z_0. The origin of the frame is defined by the user. The z_0 axis is parallel to the gravity vector, however in the opposite direction. Second is the base coordinate frame x_1–y_1–z_1, whose origin is defined by the manufacturer. Its axes are aligned with the base segment of the robot.

M. Mihelj et al., *Robotics*, https://doi.org/10.1007/978-3-319-72911-4_15

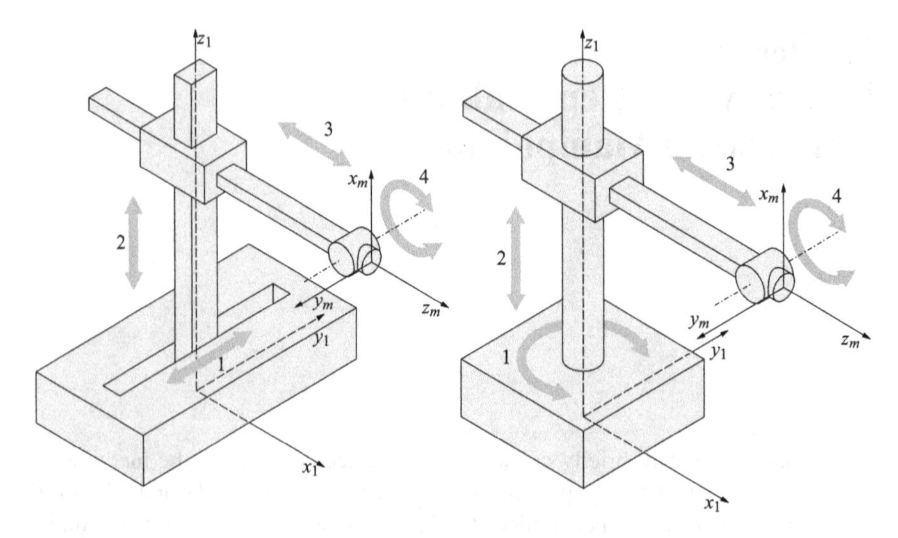

Fig. 15.1 Mechanical structures of the cartesian robot (left) and the cylindrical robot (right)

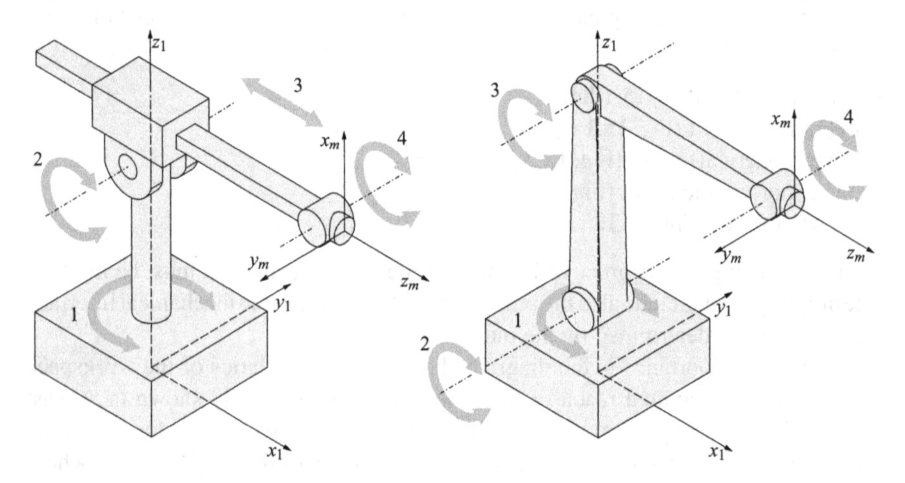

Fig. 15.2 Mechanical structure of the polar robot (left) and the anthropomorphic robot (right)

The positive z_1 axis is pointing perpendicularly away from the base mounting surface. The x_1 axis passes through the projection of the center of the robot workspace c_w. The frame x_m–y_m–z_m is called the mechanical interface coordinate frame. Its origin is placed in the center of the mechanical interface (robot palm) connecting the robot arm with the gripper. The positive z_m axis points away from the mechanical interface toward the end-effector. The x_m axis is located in the plane defined by the interface, which is perpendicular to the z_m axis.

The positive directions of robot motions, specified as the translational and rotational displacements are shown in Fig. 15.6.

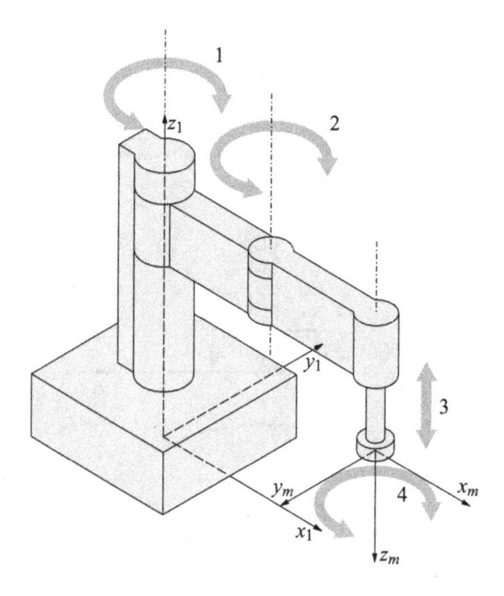

Fig. 15.3 Mechanical structure of the SCARA robot

The ISO 9283 standard deals with criteria and methods for testing of industrial robot manipulators. This is the most important standard as it facilitates the dialogue between manufacturers and users of the robot systems. It defines the way by which particular performance characteristics of a robot manipulator should be tested. The tests can be performed during the robot acceptance phase or in various periods of robot usage in order to check the accuracy and repeatability of the robot motions. The robot characteristics, which significantly affect the performance of a robot task, are the following:

- pose accuracy and repeatability (pose is defined as position, and orientation of a particular robot segment, usually end-effector),
- distance accuracy and repeatability,
- pose stabilization time,
- pose overshoot,
- drift of the pose accuracy and repeatability.

These performance parameters are important in the point-to-point robot tasks. Similar parameters are defined for cases when the robot end-effector moves along a continuous path. These parameters will not be considered in this book and can be found in the original documents.

When testing the accuracy and repeatability of a robot mechanism, two terms are important, namely the cluster and the cluster barycenter. The cluster is defined as a set of attained end-effector poses, corresponding to the same command pose. The barycenter is a point whose coordinates are the mean values of the x, y and z coordinates of all the points in the cluster. The measured position and orientation data must be expressed in a coordinate frame parallel to the base frame.

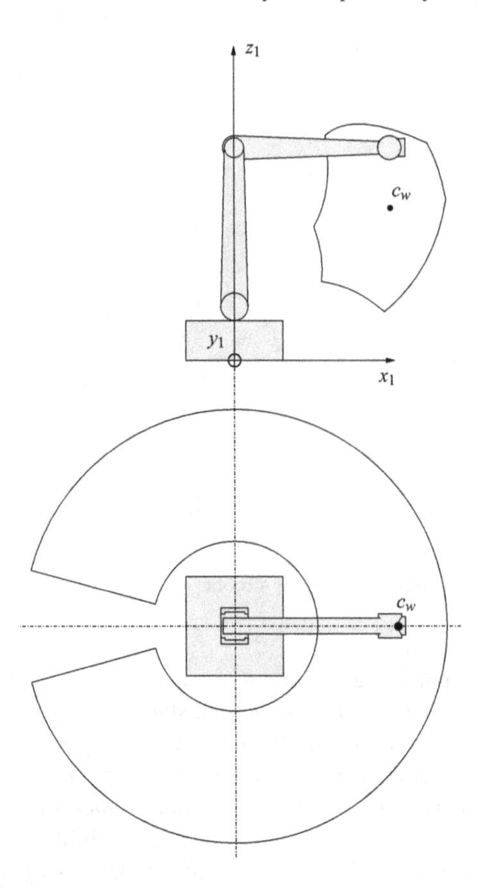

Fig. 15.4 Robot workspace

The measurement point should lie as close as possible to the origin of the mechanical interface frame. Contact-less optical measuring methods are recommended. The measuring instrumentation must be adequately calibrated. The robot accuracy and repeatability tests must be performed with maximal load at the end-effector and maximal velocity between the specified points.

The standard defines the poses which should be tested. The measurements must be performed at five points, located in a plane which is placed diagonally inside a cube (Fig. 15.7). Also specified is the pose of the cube in the robot workspace. It should be located in that portion of the workspace where most of the robot activities are anticipated. The cube must have maximal allowable volume in the robot workspace and its edges should be parallel to the base coordinate frame. The point P_1 is located in the intersection of the diagonals in the center of the cube. The points $P_2 - P_5$ are located at a distance from the corners of the cube equal to $10\% \pm 2\%$ of the length of the diagonal L. The standard also determines the minimum number of cycles to be performed when testing each characteristic parameter:

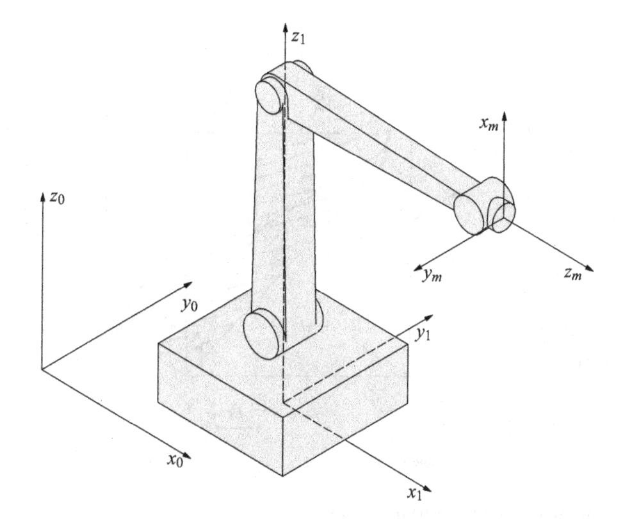

Fig. 15.5 The coordinate frames of the robot manipulator

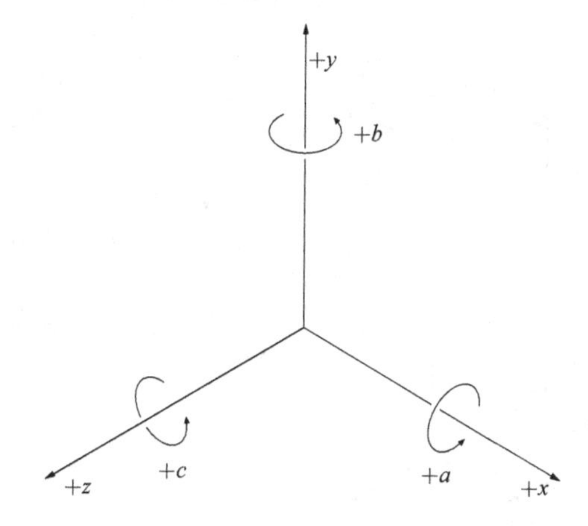

Fig. 15.6 Positive directions of translational and rotational displacements

- pose accuracy and repeatability: 30 cycles,
- distance accuracy and repeatability: 30 cycles,
- pose stabilization time: 3 cycles,
- pose overshoot: 3 cycles,
- drift of pose accuracy and repeatability: continuous cycling during 8 h.

When testing the accuracy and repeatability of the end-effector poses we must distinguish between the so-called command pose and the attained pose (Fig. 15.8).

The command pose is the desired pose, specified through robot programming or manual input of the desired coordinates using a teach pendant. The attained pose is the actually achieved pose of the robot end-effector in response to the command

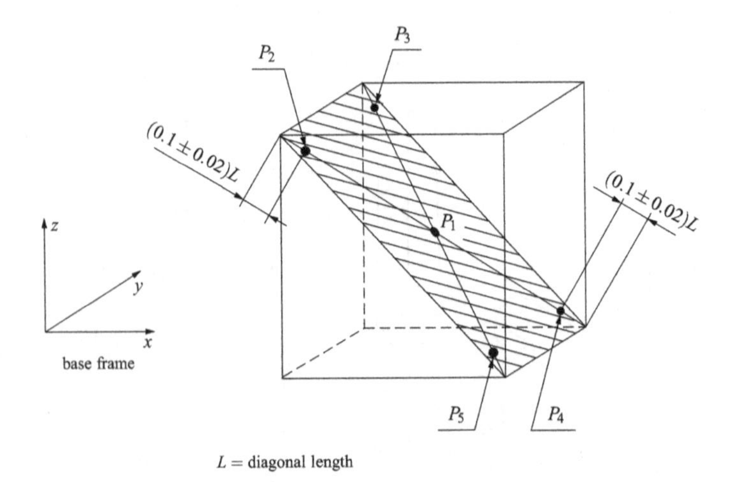

Fig. 15.7 The cube with the points to be tested

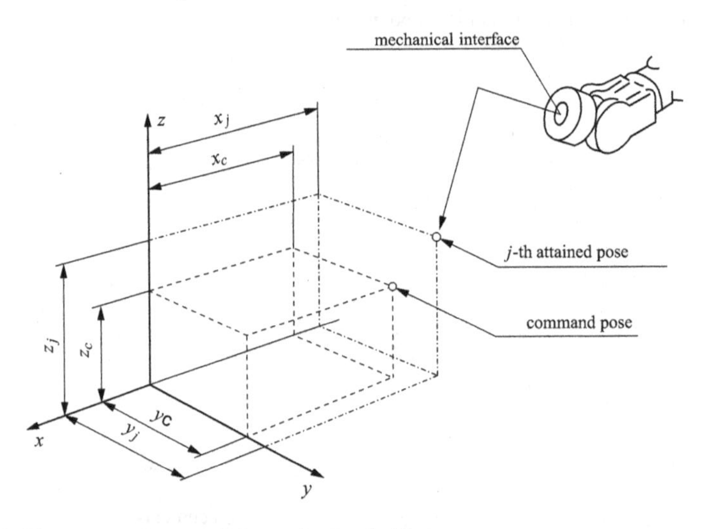

Fig. 15.8 The command pose and the attained end-effector pose

pose. The pose accuracy evaluates the deviations, which occur between the command and the attained pose. The pose repeatability estimates the fluctuations in the attained poses for a series of repeated visits to the same command pose. The pose accuracy and repeatability are, therefore, very similar to the accuracy and repeatability of repetitive shooting at a target. The reasons for the deviations are: errors caused by the control algorithm, coordinate transformation errors, differences between the dimensions of the robot mechanical structure and the robot control model, mechanical faults, such as hysteresis or friction, and external influences such as temperature.

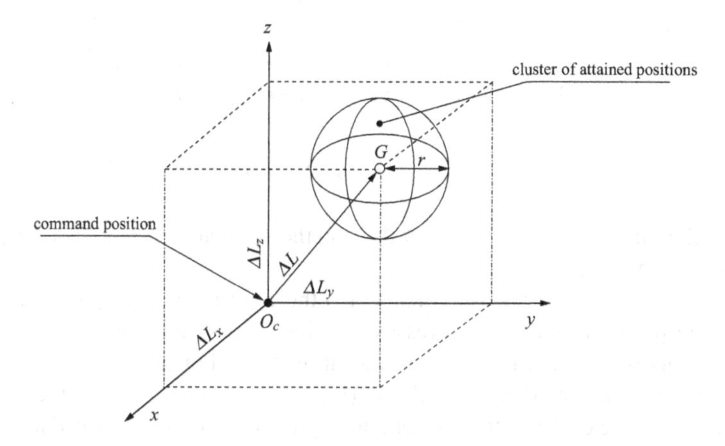

Fig. 15.9 The position accuracy and repeatability

The pose accuracy is defined as the deviation between the command pose and the mean value of the attained poses when the end-effector was approaching the command pose from the same direction. The position and orientation accuracy are treated separately. The position accuracy is determined by the distance between the command pose and the barycenter of the cluster of attained poses (Fig. 15.9). The position accuracy $\Delta L = [\Delta L_x \ \Delta L_y \ \Delta L_z]^T$ is expressed by the following equation

$$\Delta L = \sqrt{(\bar{x} - x_c)^2 + (\bar{y} - y_c)^2 + (\bar{z} - z_c)^2}, \tag{15.1}$$

where $(\bar{x}, \bar{y}, \bar{z})$ are the coordinates of the cluster barycenter, obtained by averaging the 30 measurement points, assessed when repeating the motions into the same command pose O_c with the coordinates (x_c, y_c, z_c).

The orientation accuracy is the difference between the commanded angular orientation and the average of the attained angular orientations. It is expressed separately for each axis of the base coordinate frame. The orientation accuracy around the z axis has the following form

$$\Delta L_c = \bar{C} - C_c, \tag{15.2}$$

where $\bar{C}$ is the mean value of the orientation angles around the z axis, obtained in 30 measurements when trying to reach the same command angle C_c. Similar equations are written for the orientation accuracy around the x and y axes.

The standard exactly defines also the course of the measurements. The robot starts from point P_1 and moves into points P_5, P_4, P_3, P_2, P_1. Each point is always reached from the same direction

$$\begin{array}{ll} 0 \text{ cycle} & \quad\quad\quad\quad\quad\; P_1 \\ 1^{st} \text{ cycle} & P_5 \rightarrow P_4 \rightarrow P_3 \rightarrow P_2 \rightarrow P_1 \\ 2^{nd} \text{ cycle} & P_5 \rightarrow P_4 \rightarrow P_3 \rightarrow P_2 \rightarrow P_1 \\ & \quad\quad\quad\quad\quad\; \vdots \\ 30^{th} \text{ cycle} & P_5 \rightarrow P_4 \rightarrow P_3 \rightarrow P_2 \rightarrow P_1 \end{array}$$

For each point the position accuracy ΔL and the orientation accuracies ΔL_a, ΔL_b and ΔL_c are calculated.

For the same series of measurements also the pose repeatability is to be determined. The pose repeatability expresses the closeness of the positions and orientations of the 30 attained poses when repeating the robot motions into the same command pose. The position repeatability (Fig. 15.9) is determined by the radius of the sphere r whose center is the cluster barycenter. The radius is defined as

$$r = \bar{D} + 3S_D. \tag{15.3}$$

The calculation of the radius r according Eq. (15.3) is further explained by the following equations

$$\begin{aligned} \bar{D} &= \frac{1}{n}\sum_{j=1}^{n} D_j \\ D_j &= \sqrt{(x_j - \bar{x})^2 + (y_j - \bar{y})^2 + (z_j - \bar{z})^2} \\ S_D &= \sqrt{\frac{\sum_{j=1}^{n}(D_j - \bar{D})^2}{n-1}}. \end{aligned} \tag{15.4}$$

In the above equations we again select $n = 30$, while (x_j, y_j, z_j) are the coordinates of the j-th attained position.

The orientation repeatability for the angle around the z axis is presented in Fig. 15.10. The orientation repeatability expresses how dispersed are the 30 attained angles around their average for the same command angle. It is described by the threefold standard deviations. For the angle around the z axis we have

$$r_c = \pm 3S_c = \pm 3\sqrt{\frac{\sum_{j=1}^{n}(C_j - \bar{C})^2}{n-1}}. \tag{15.5}$$

In Eq. (15.5) C_j represents the angle measured at the j-th attained pose. The course of the measurements is the same as in testing of the accuracy. The radius r and the angular deviations r_a, r_b and r_c are calculated for each pose separately.

The distance accuracy and repeatability are tested in a similar way. The distance accuracy quantifies the deviations which occur in the distance between two command positions and two sets of the mean attained positions. The distance repeatability

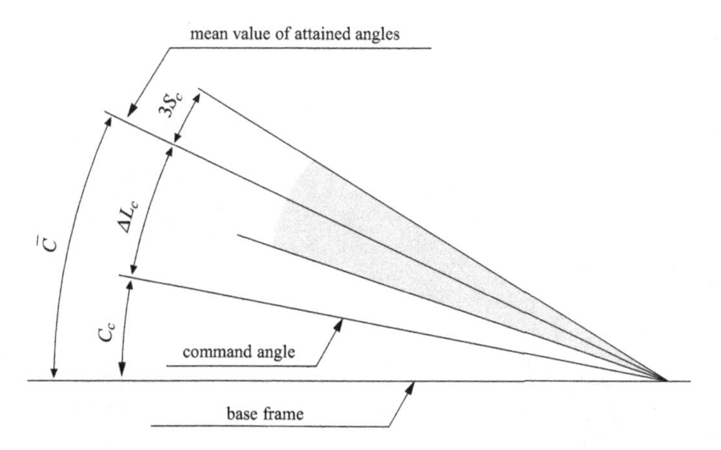

Fig. 15.10 The orientation accuracy and repeatability

determines the fluctuations in distances for a series of repeated robot motions between two selected points. The distance accuracy is defined as the deviation between the command distance and the mean of the attained distances (Fig. 15.11). Assuming that P_{c1} and P_{c2} are the commanded pair of positions and P_{1j} and P_{2j} are the j-th pair from the 30 pairs of the attained positions, the distance accuracy ΔB is defined as

$$\Delta B = D_c - \bar{D}. \tag{15.6}$$

where

$$D_c = |P_{c_1} - P_{c_2}| = \sqrt{(x_{c_1} - x_{c_2})^2 + (y_{c_1} - y_{c_2})^2 + (z_{c_1} - z_{c_2})^2}$$

$$\bar{D} = \frac{1}{n}\sum_{j=1}^{n} D_j$$

$$D_j = |P_{1_j} - P_{2_j}| = \sqrt{(x_{1_j} - x_{2_j})^2 + (y_{1_j} - y_{2_j})^2 + (z_{1_j} - z_{2_j})^2}.$$

In the above equations describing the distance accuracy $P_{c_1} = (x_{c_1}, y_{c_1}, z_{c_1})$ and $P_{c_2} = (x_{c_2}, y_{c_2}, z_{c_2})$ represent the pair of desired positions while $P_{1_j} = (x_{1_j}, y_{1_j}, z_{1_j})$ and $P_{2_j} = (x_{2_j}, y_{2_j}, z_{2_j})$ are the pair of attained positions. The distance accuracy test is performed at maximal loading of the robot end-effector, which must be displaced 30 times between points P_2 and P_4 of the measuring cube. The distance repeatability R_B is defined as

$$R_B = \pm 3\sqrt{\frac{\sum_{j=1}^{n}(D_j - \bar{D})^2}{n-1}}. \tag{15.7}$$

Let us consider another four characteristic parameters which should be tested in industrial robots moving from point to point. The first is the pose stabilization

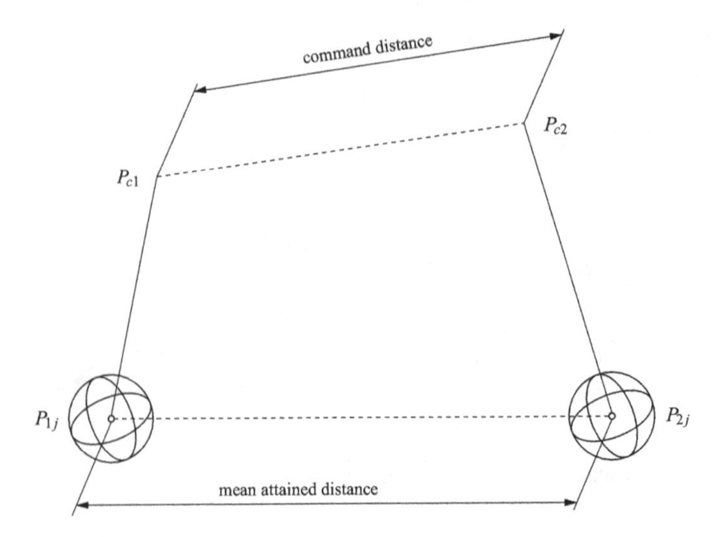

Fig. 15.11 Distance accuracy

time. The stabilization time is the time interval between the instant when the robot gives the “attained pose” signal and the instant when either oscillatory or damped motion of the robot end-effector falls within a limit specified by the manufacturer. The definition of the pose stabilization time is evident from Fig. 15.12. The test is performed at maximal loading and velocity. All five measuring points are visited in the following order $P_1 \rightarrow P_2 \rightarrow P_3 \rightarrow P_4 \rightarrow P_5$. For each pose the mean value of three cycles is calculated.

A similar parameter is the pose overshoot, also shown in Fig. 15.12. The pose overshoot is the maximum deviation between the approaching end-effector trajectory and the attained pose after the robot has given the “attained pose” signal. In Fig. 15.12 a negative overshoot is shown in the first example and a positive overshoot in the second example. The instant $t = 0$ is the time when the “attained pose” signal was delivered. The measuring conditions are the same as when testing the stabilization time.

The last two parameters to be tested in the industrial robot manipulator moving from point to point are drift of the pose accuracy and the drift of the pose repeatability. The drift of the position accuracy L_{DR} is defined as

$$L_{DR} = |\Delta L_{t=0} - \Delta L_{t=T}|, \tag{15.8}$$

where $\Delta L_{t=0}$ and $\Delta L_{t=T}$ are the position accuracy values at time $t = 0$ and time t=T, respectively. The drift of the orientation accuracy L_{DRC} is equal to

$$L_{DRC} = \left|\Delta L_{c,t=0} - \Delta L_{c,t=T}\right|, \tag{15.9}$$

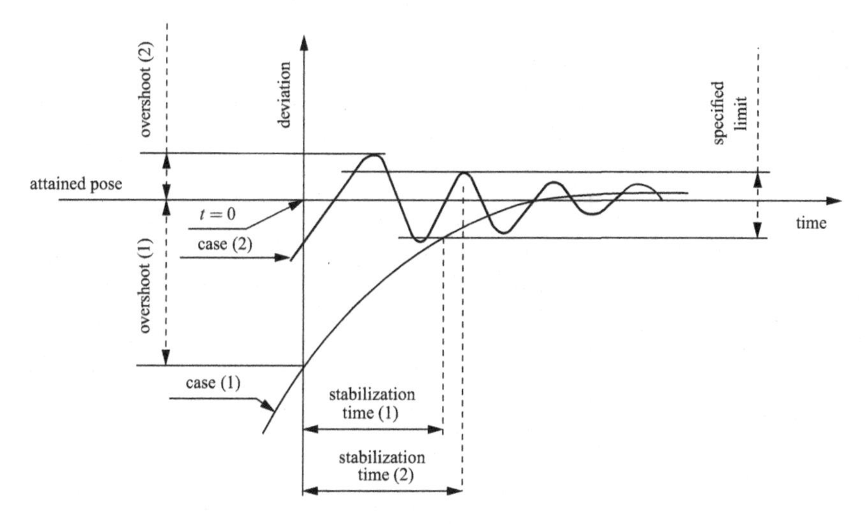

Fig. 15.12 Pose stabilization time and overshoot

where $\Delta L_{c,t=0}$ and $\Delta L_{c,t=T}$ are the orientation accuracy values at time $t = 0$ and time t=T, respectively. The drift of the position repeatability is defined by the following equation

$$r_{DR} = r_{t=0} - r_{t=T}, \tag{15.10}$$

where $r_{t=0}$ and $r_{t=T}$ are the position repeatability values at time $t = 0$ and time t=T, respectively. The drift of the orientation repeatability is for the rotation around the z axis defined as

$$r_{DRC} = r_{c,t=0} - r_{c,t=T}, \tag{15.11}$$

where $r_{c,t=0}$ and $r_{c,t=T}$ are the orientation repeatability values at time $t = 0$ and time t=T, respectively. The measurements are performed at maximal robot loading and velocity. The robot is cyclically displaced between points P_4 and P_2. The cyclic motions last for eight hours. Measurements are only taken in point P_4.

Fig. 15.12 [illegible] overlap

[illegible] M_{en} [illegible] and M [illegible] the [illegible] accuracy [illegible] [illegible] the [illegible] is [illegible] the following [illegible]

$$[illegible]$$ (15.4)

where [illegible] and [illegible] are [illegible] respectively [illegible]

$$[illegible]$$ (15.5)

where [illegible] [illegible] [illegible] [illegible] points P and P [illegible] in point [illegible]

Appendix A
Derivation of the Acceleration in Circular Motion

Let us first recall the definitions of position vector, velocity and acceleration of a particle. In a given reference frame (i.e., coordinate system) the position of a particle is given by a vector extending from the coordinate frame origin to the instantaneous position of the particle. This vector could be a function of time, thus specifying the particle trajectory

$$\mathbf{r}(t) = (x(t), y(t), z(t)). \tag{A.1}$$

The velocity of the particle is defined as the change of position per unit time

$$\mathbf{v} = \lim_{\Delta t \to 0} \frac{\Delta \mathbf{r}}{\Delta t} = \frac{d\mathbf{r}}{dt}. \tag{A.2}$$

The acceleration is defined as the change of velocity per unit time,

$$\mathbf{a} = \lim_{\Delta t \to 0} \frac{\Delta \mathbf{v}}{\Delta t} = \frac{d\mathbf{v}}{dt}. \tag{A.3}$$

We note that this is a vector equation, so the change in velocity refers to both a change in the magnitude of velocity, as well as to a change of velocity direction.

Circular motion is described by a rotating vector $\mathbf{r}(t)$ of fixed length, $|r| = constant$. The position vector is thus determined by the radius of the circle r, and by the angle $\theta(t)$ of $\mathbf{r}$ with respect to the x-axis (Fig. A.1).

Let us now introduce a set of three orthogonal unit vectors: $\mathbf{e}_r$ in the direction of $\mathbf{r}$, $\mathbf{e}_t$ in the direction of the tangent to the circle and $\mathbf{e}_z$ in the direction of the z axis. The relation between the three unit vectors is given by $\mathbf{e}_t = \mathbf{e}_z \times \mathbf{e}_r$.

We define the angular velocity vector as being perpendicular to the plane of the circular trajectory with magnitude equal to the time derivative of the angle θ

$$\omega = \dot{\theta}\mathbf{e}_z. \tag{A.4}$$

Let us proceed to calculate the velocity

M. Mihelj et al., *Robotics*, https://doi.org/10.1007/978-3-319-72911-4

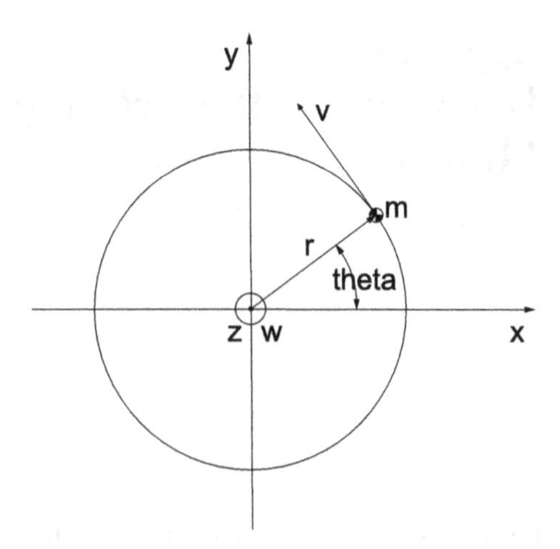

Fig. A.1 Parameters and variables in circular motion

$$\mathbf{v} = \frac{d\mathbf{r}}{dt}. \tag{A.5}$$

The direction of velocity is given by the tangent to the circle: $\mathbf{e}_t = \mathbf{e}_z \times \mathbf{e}_r$. The magnitude of velocity is given by the length of the infinitesimal circular arc $ds = rd\theta$ divided by the infinitesimal time dt, that the particle requires to traverse this path

$$\frac{ds}{dt} = r\frac{d\theta}{dt} = r\dot{\theta}. \tag{A.6}$$

Including the tangential direction of velocity gives

$$\mathbf{v} = r\dot{\theta}\mathbf{e}_t = \dot{\theta}\mathbf{e}_z \times r\mathbf{e}_r = \boldsymbol{\omega} \times \mathbf{r}. \tag{A.7}$$

In order to obtain the acceleration, we calculate the time derivative of velocity

$$\mathbf{a} = \frac{d\mathbf{v}}{dt} = \frac{d}{dt}(\boldsymbol{\omega} \times \mathbf{r}). \tag{A.8}$$

We differentiate the vector product as one would differentiate a normal product of two functions

$$\mathbf{a} = \frac{d\boldsymbol{\omega}}{dt} \times \mathbf{r} + \boldsymbol{\omega} \times \frac{d\mathbf{r}}{dt}. \tag{A.9}$$

Defining the angular acceleration $\boldsymbol{\alpha}$ as the time derivative of angular velocity $\boldsymbol{\alpha} = \frac{d\boldsymbol{\omega}}{dt}$, we see that the first term corresponds to the tangential acceleration

$$\mathbf{a}_t = \boldsymbol{\alpha} \times \mathbf{r}. \tag{A.10}$$

In the second term we insert the expression for velocity

$$\frac{d\mathbf{r}}{dt} = \mathbf{v} = \omega \times \mathbf{r}, \tag{A.11}$$

and we get a double vector product $\omega \times (\omega \times \mathbf{r})$. Using the identity $\mathbf{a} \times (\mathbf{b} \times \mathbf{c}) = \mathbf{b}(\mathbf{a} \cdot \mathbf{c}) - \mathbf{c}(\mathbf{a} \cdot \mathbf{b})$ from vector algebra and noting that ω and $\mathbf{r}$ are orthogonal, we obtain for the second term in the equation for acceleration

$$\omega \times \frac{d\mathbf{r}}{dt} = \omega \times (\omega \times \mathbf{r}) = \omega(\omega \cdot \mathbf{r}) - \mathbf{r}(\omega \cdot \omega) = -\omega^2\mathbf{r}, \tag{A.12}$$

which is the radial (or centripetal) component of acceleration. So we finally have

$$\mathbf{a} = \mathbf{a}_t + \mathbf{a}_r = \alpha \times \mathbf{r} - \omega^2\mathbf{r}. \tag{A.13}$$

Further Reading

1. Bajd T, Mihelj M, Munih M (2013) Introduction to Robotics, Springer
2. Craig JJ (2005) Introduction to Robotics—Mechanics and Control, Pearson Prentice Hall
3. Kajita S, Hirukawa H, Harada K, Yokoi K (2014) Introduction to Humanoid Robotics, Springer
4. Klančar G, Zdešar A, Blažič S, Škrjanc I (2017) Wheeled Mobile Robotics - From Fundamentals Towards Autonomous Systems, Elsevier
5. Lenarčič J, Bajd T, Stanišić MM (2013) Robot Mechanisms, Springer
6. Merlet J-P (2006) Parallel Robots (Second Edition), Springer
7. Mihelj M, Podobnik J (2012) Haptics for Virtual Reality and Teleoperation, Springer
8. Mihelj M, Novak D, Beguš S (2014) Virtual Reality Technology and Applications, Springer
9. Natale C (2003) Interaction Control of Robot Manipulators, Springer
10. Nof SY (1999) Handbook of Industrial Robotics, John Wiley & Sons
11. Paul RP (1981) Robot Manipulators: Mathematics, Programming, and Control, The MIT Press
12. Sciavico L, Siciliano B (2002) Modeling and Control of Robot Manipulators, Springer
13. Spong MW, Hutchinson S, Vidyasagar M (2006) Robot Modeling and Control, John Wiley & Sons
14. Tsai LW (1999) Robot Analysis: The Mechanics of Serial and Parallel Manipulators, John Wiley & Sons
15. Xie M (2003) Fundamentals of Robotics—Linking Perception to Action, World Scientific

Index

M. Mihelj et al., *Robotics*, https://doi.org/10.1007/978-3-319-72911-4